Clients Driving Innovation

Clients Driving Innovation

Edited by

Peter Brandon
Director of Salford University Think Lab and Director Strategic Programmes
School of the Built Environment
University of Salford, UK

Shu-Ling Lu
Research Fellow
School of the Built Environment
University of Salford, UK

THINKlab

⊛WILEY-BLACKWELL

A John Wiley & Sons, Ltd., Publication

This edition first published 2008
© 2008 Blackwell Publishing Ltd

Blackwell Publishing was acquired by John Wiley & Sons in February 2007. Blackwell's publishing programme has been merged with Wiley's global Scientific, Technical, and Medical business to form Wiley-Blackwell.

Registered office
John Wiley & Sons Ltd, The Atrium, Southern Gate, Chichester, West Sussex, PO19 8SQ, United Kingdom

Editorial offices
9600 Garsington Road, Oxford, OX4 2DQ, United Kingdom
350 Main Street, MA 02148-5020, USA

For details of our global editorial offices, for customer services and for information about how to apply for permission to reuse the copyright material in this book please see our website at www.wiley.com/wiley-blackwell.

The right of the author to be identified as the author of this work has been asserted in accordance with the Copyright, Designs and Patents Act 1988.

Wiley also publishes its books in a variety of electronic formats. Some content that appears in print may not be available in electronic books.

Designations used by companies to distinguish their products are often claimed as trademarks. All brand names and product names used in this book are trade names, service marks, trademarks or registered trademarks of their respective owners. The publisher is not associated with any product or vendor mentioned in this book. This publication is designed to provide accurate and authoritative information in regard to the subject matter covered. It is sold on the understanding that the publisher is not engaged in rendering professional services. If professional advice or other expert assistance is required, the services of a competent professional should be sought.

Library of Congress Cataloging-in-Publication Data

Clients driving innovation / edited by Peter Brandon, Shu-Ling Lu.
    p.   cm.
  Includes bibliographical references and index.
  ISBN-13: 978-1-4051-7566-1 (hardback : alk. paper)
  ISBN-10: 1-4051-7566-4 (hardback : alk. paper)
    1. Building–Technological innovations.   2. Construction industry–Customer services.
3. Customer relations.   4. Consumer satisfaction.   5. Production planning.
I. Brandon, P. S. (Peter S.)   II. Lu, Shu-Ling.

TH153.C577   2008
690.068'8–dc22
                        2008002543

A catalogue record for this book is available from the British Library.

Set in 9.5/12 pt Palatino by Aptara Inc., New Delhi, India
Printed in Singapore by Markono Print Media Pte Ltd

1 2008

Contents

Note on editors

Professor Peter Brandon DSc, DEng, MSc (Arch), FRICS, ASAQS Director, Salford University Think Lab and Director, Strategic Programmes, School of the Built Environment, University of Salford, UK.

Professor Brandon is a former Pro-Vice-Chancellor for research at the University of Salford, the only UK University to be awarded a 6-star rating in the Built Environment within the independent UK research assessment exercise. His research interests range across construction economics and management, information and knowledge-based systems for construction and more recently sustainable development. He has published widely including 18 books as author, co-author and editor plus over 250 papers worldwide. Several of the outputs of his research have resulted in commercial projects.

He has played a significant role in UK Construction Research Policy including serving as Chairman of the UK Science and Engineering Research Council Panel for Construction and Chairman of the UK Research Assessment Exercise Panel for the Built Environment (1996 and 2001).

Dr. Shu-Ling Lu PhD, MSc, BSc, Dip (Arch) Research Fellow, School of the Built Environment, University of Salford, UK.

Doctor Shu-Ling Lu PG Cert a senior researcher within the Research Institute of the Built and Human Environment at the University of Salford in the UK. She is the Joint Co-ordinator of the International Council for Research and Innovation in Building and Construction (CIB) Task Group 65 in the Management of Small Construction Firms.

Dr Lu's main research area includes innovation management within small construction firms (particularly within knowledge-intensive professional service firms), gender issues in construction and academia-industry engagement. Dr Lu has published 1 book, 2 book chapters, and 40 journal and conference papers. Dr Lu has been invited to provide a number of keynote addresses in the areas of knowledge and quality management.

Contributors

Carl Abbott
Manager, Salford Centre for Research and Innovation (SCRI), University of Salford, Salford, UK

Brian Atkin
Visiting Professor, Department of Construction Management, Lund University, Lund, Sweden

Peter Barrett
Professor, Construction and Property Management, Pro-Vice-Chancellor, Research and Graduate Studies, University of Salford, UK and President, International Council for Research and Innovation in Building and Construction, Rotterdam, The Netherlands

Frédéric Bougrain
Doctor, Department of Economics and Social Sciences, Centre Scientifique et Technique du Bâtiment, Paris, France

Mike Bresnen
Professor, Organisational Behaviour, School of Management, University of Leicester, Leicester, UK

Jan Bröchner
Professor, Organization of Construction, Department of Technology Management and Economics, Chalmers University of Technology, Göteborg, Sweden

William L. Cate
The Cate Group, Miami, FL, USA

Roger Courtney
Professorial Fellow, Construction Innovation, University of Manchester, Manchester, UK

Mohammed F. Dulaimi
Doctor, Construction Management and Innovation, Institute of Engineering, The British University in Dubai, Dubai, United Arab Emirates

Erica Dyson
Visiting Professor, Healthcare and Regeneration, School of the Built Environment, University of Salford, Salford, UK and Director, Development and Redesign, Trafford Healthcare NHS Trust, Manchester, UK

Charles Egbu
Professor, Project Management and Strategic Management in Construction, School of the Built Environment, University of Salford, Salford, UK

Eileen Fairhurst
Professor, Health and Ageing Policy Studies, Manchester Metropolitan University, Manchester, UK and Chairman, Salford Teaching Primary Care Trust, Salford, UK

Aminah Robinson Fayek
Professor, NSERC/Alberta Construction Industry Associate Research Chair in Construction Engineering and Management, Department of Civil and Environmental Engineering, University of Alberta, Edmonton, Alberta, Canada

Patrick S. W. Fong
Associate Professor, Department of Building and Real Estate, The Hong Kong Polytechnic University, Kowloon City, Hong Kong

Nuno Gil
Senior Lecturer, Deputy Director, Centre for Research in the Management of Projects (CRMP), Manchester Business School, University of Manchester, Manchester, UK

Colin Gray
Professor, Management and Production Engineering, Academic Director, Health and Care Infrastructure Research and Innovation Centre (HaCIRIC, Reading Team), School of Construction Management and Engineering, University of Reading, Reading, UK

Päivi Haapalainen
PhD Researcher, Production Department, University of Vaasa, Vaasa, Finland

Andreas Hartmann
Assistant Professor, Department of Construction Management and Engineering, University of Twente, Twente, The Netherlands

John Hobson
Visiting Professor, School of the Built Environment, University of Salford, Salford, UK and Independent policy analyst, Construction Director (Formerly), Department of Trade and Industry, London, UK

Leif Hommen
Associate Professor, Research and Competence in the Learning Economy (CIRCLE), Lund University, Lund, Sweden

Mike Kagioglou
Professor, Process Management, Co-Director, Salford Centre for Research and Innovation (SCRI), University of Salford, Salford, UK and Academic Director, Health and Care Infrastructure Research and Innovation Centre (HaCIRIC), University of Salford, Salford, UK

John M. Kamara
Senior Lecturer, Director of Postgraduate Research, Coordinator, Applied Research in Architecture (ARA) Group, School of Architecture, Planning and Landscape, Newcastle University, Newcastle upon Tyne, UK

Robert Kay
Head, Strategic Innovation, Westpac Banking Corporation, Sydney, Australia

Heikki Lonka
Deputy Mayor, Vaasa town, Finland

Shu-Ling Lu
Research Fellow, School of the Built Environment, University of Salford, Salford, UK

Roger Miller
Jarislowsky Professor, Innovation and Project Management, Department of Mathematics and Industrial Engineering, École Polytechnique, Montreal, Québec, Canada

Marcela Miozzo
Professor, Economics and Management of Innovation, Manchester Business School, University of Manchester, Manchester, UK

Marja Naaranoja
Director, Masters Program in Construction Engineering, VAMK, University of Applied Sciences, Vaasa, Finland and Adjunct Professor, University of Vaasa, Vaasa, Finland

Jeff H. Rankin
Associate Professor, M. Patrick Gillin Chair in Construction Engineering and Management, Department of Civil Engineering, University of New Brunswick, Saint John, New Brunswick, Canada

Martin Sexton
Professor, Construction Management, School of the Built Environment, University of Salford, Salford, UK

E. Sarah Slaughter
Senior Lecturer, Sloan School of Management, Massachusetts Institute of Technology, Cambridge, MA, USA

Kenneth Treadaway
Visiting Professor, School of the Built Environment, University of Salford, Salford, UK

Ernie Tromposch
Program Leader, Construction Management, Project Management Office, Nova Chemicals Corporation, Calgary, Alberta, Canada

Patricia Tzortzopoulos
Academic Fellow, School of the Built Environment, University of Salford, Salford, UK

Kristian Widén
Assistant Professor, Division of Construction Management, Lund University, Lund, Sweden

Graham M. Winch
Professor, Project Management, Director, Centre for Research in the Management of Projects, Manchester Business School, University of Manchester, Manchester, UK

Chris Woods
Professor, R&D, Wates Group Ltd, Surrey, UK and Visiting Professor, School of the Built Environment, University of Salford, Salford, UK

Note on CIB

CIB – the International Council for Research and Innovation in Building and Construction – is an association that provides a worldwide network for exchange concerning all aspects of buildings and the built environment during all stages of their life cycle. CIB Members are companies, organisations and individuals active in the research community, industry, government and education who cooperate in a programme of over fifty scientific commissions. This book is an outcome from the work of CIB Task Group TG85 – Clients and Construction Innovation.

Note on Think Lab

THINKlab

This book arises from debate within the internationally leading University of Salford 'Think Lab'. This state-of-the-art facility has been developed for research into Information and Communication Technologies (ICTs) in many fields including design and construction. It provides a forum for leading figures across the world to participate, both in person and through virtual collaborative technologies, to discuss topics relating to future developments in ICTs applied to various topic areas. For further information visit www.thinklab.salford.ac.uk.

Acknowledgements

First and foremost, the editors of this book would like to thank all the authors for sharing their knowledge and insight. Without their support, this book could not have been produced.

We received great encouragement from the International Council for Research and Innovation in Building and Construction (CIB) for the organisation of the 'Clients Driving Innovation' workshop (which initiated and then provided the motivation for the creation of this book) and we would like to acknowledge their prominent role.

We would like to thank and acknowledge the valuable assistance of Hanneke van Dijk for her expertise in the managing of correspondence and for her time given to proof-reading. Her enthusiasm and commitment have proved invaluable.

The editors and publisher gratefully acknowledge those who have granted permission to reproduce material in this book. Although every effort has been made to secure permission to publish prior to publication, we take this opportunity to offer our apologies for any errors and omissions. If notified, we will endeavour to correct these at the earliest opportunity.

Preface – Clients driving innovation?

Peter Brandon

The role of the client in driving innovation

In recent years the construction industry, and the professions associated with the built environment, have been criticised for their lack of innovation compared with the revolutionary developments that have been seen in many other major industries. This is, of course, a relative judgement as the industry has indeed innovated and evolved over many centuries from the time when human kind decided to create its own shelter. The dependence on the materials derived from the land, whether renewable or not, meant that the industry was largely local and regional and its development depended on craft processes handed down from generation to generation. The degree of innovation was limited by the nature of the labour skills, technology and materials that were available. Other manufacturing industries are a more recent phenomenon, have tended to be global, and have been forged from a strong technical base that in the last century has required a rationalisation of the process supported by technical development to remain competitive. It appears that construction has not previously had to respond to these pressures.

Nevertheless, the question of why construction has not been seen to innovate to the same extent is being raised in many quarters across the world. There has even been a book written that asks the question 'Why is Construction so Backward?' (Woudhuysen and Abley, 2004). This has created concern among many involved with construction and property as to where should the motivation and drive for innovation in one of the world's largest industries come from? In other industries, it seems that it is the competitive nature of the market that has driven firms to find new solutions to the problems faced by all those engaged, from the clients to the professional consultants to the contractors through to the supply chain. In fact, many of them have looked to changing the process to make sure that they remain competitive in the market that they address. Construction has remained stubbornly immune from these pressures possibly because of the localisation of its markets until comparatively recently. A change has occurred that may be the result of the growing internationalisation of the construction firms (at the time of writing six of the largest construction firms in the UK are foreign owned) whereby the firm has to compete in a faster moving market in which the supply chains may be stretched across the world. It may also be a function of the changes in corporate leadership whereby chairmen and chief executives may come from other industries and find the construction sector rather primitive in its approach to the process it is trying to enact. The prime example in the UK would be Sir John Egan, who came from Jaguar Cars to take the Chair of British Airports Authority (BAA) and who then led a client drive to improve construction industry performance. He started with his own company and then extended the principles to the rest of the industry culminating in a major report (Egan, 1998) that led to considerable new thinking within the construction team. It will be interesting to see whether these ideas continue to develop without the government funding that encouraged these innovative approaches.

Client innovation – a challenge to the industry?

Whilst it can be said that clients have made an impact in this industry and there are clear examples in this book, the industry itself seems divided as to whether it is the role of clients to drive innovation. On the one side, there is the argument that only the client *knows* what innovation he or she requires and often he or she is the only person able to take an overview across all aspects of the process from inception through design, assembly and then occupation of the final artefact. Since much of the innovation is likely to come from the integration of processes then who else, it is argued, can have the vision for change and encourage innovation across all the actors in the process. On the other side, members of the industry and others query whether this is just a failure of the industry itself to resolve its own problems. What other industry, it is argued, demands that the client take the leading role. If firms do not innovate they die. It does not require the recipients of innovation to drive the process of beneficial change! To suggest that the client should take this role is a 'cop out' for the industry and discourages it from investing in its own development. A review of many products will show that from motor cars to electrical goods and from steel production to ship building it is the producers of the goods who create the innovation and they survive and thrive because of it. Where they do not, and you only have to look at the automobile and motor cycle manufacturers in the UK during the latter part of the 20th century, they die or are taken over by others who will innovate.

This raises, of course, all sorts of interesting questions as to why innovation has not been a high priority in the construction and property industry. Many different reasons have been given including the following:

- The structure of the industry (and particularly the large number of small firms) militates against change as none can give the time and investment required to change not only their own practice but also the change in others to which they relate. There are a huge number of interfaces in construction, each of which discourages change. However, other industries such as the aircraft industries have similar structures but can change process to suit.
- The education of a large proportion of the industry outside the professions is not sufficient to take up new technologies. This may be true but it may mean shedding low-level labour and replacement by machine as has happened elsewhere.
- The local nature of the industry markets prevents a global brand or product to be developed around which the technology can be developed. However, this is changing as more components are manufactured off-site (and off shore!) and the competition is becoming more globally competitive.
- There is no incentive for innovation. In the past, this must have been true as price became the determinant between firms and not their ability to improve the practice. However, it is increasingly true to say that innovation does create competitive advantage in aspects such as design and manufacture and this expertise is providing

a greater demand for the services that achieve it. Just look at some of the major designers around the world and the engineering firms who succeed in making new designs possible and you see the growth of firms based on their innovative methods.
- The 'lowest cost wins' approach denies the firms the capital to invest in research and innovation. There are not many industries in the world that exist on 'highest cost wins' of course! It is often argued that value is the key and not cost but innovation should try and create increased value that is a balance between cost and performance.

This book begins an investigation as to whether clients have a role in overcoming these perceptions of why the industry is not innovating at the same pace as its compatriot industries. In order to do this, it needs to call on the knowledge and experience of industrialists, clients, academics and research bodies to establish a line of enquiry that will test the thesis that clients do have a part to play but just as importantly what are the building blocks of knowledge that are necessary to build the thesis. Like all topics that emerge over time, in their inception they are largely unstructured and lack coherence. It takes time for the components to emerge in a structured way so that others can build on what has gone before and begin the painstaking business of formulating theory and practice that will lead to new insights and improved behaviour. Whilst management science has explored the role of clients in many industries and particularly manufacturing, the role of clients in construction and property is largely virgin territory.

Starting assumptions

In considering a topic as broad and unstructured as clients and innovation in the construction and property industry, it is advisable to identify the starting assumptions on which a study can take place. These include the following:

- The starting assumption must be that clients have a role to play in driving innovation but this is a hypothesis that needs to be tested in the arguments that are presented.
- There is a generally accepted definition of both client and innovation so that a useful discussion can take place based on mutual understanding.
- The structure of the industry will not change overnight and, therefore, the professional boundaries, size of firms and other attributes that we take for granted may change in the longer term and be encouraged by the innovation but they provide the framework within which we work at the present time.
- Risk distribution is fair to all parties so that the innovation is not seen as something that is detrimental to the health of one sector or to the whole of the industry. However, in large scale innovation it is often found that one group can suffer at the expense of another.
- The technological, economic and cultural environment is changing rapidly and all firms and clients must adjust to this external stimulus to innovation. The industry does not exist in isolation to those it serves.

Some of the issues we need to address may seem obvious but they may be those that need challenging. Over time, we begin to accept assumptions without challenge because we have been educated to accept them, have used them successfully, and they have provided a language within which we can work with others. They provide the 'norm' that enables us to go about our daily business without resorting to the tiresome task of going back to first principles. They provide 'rules of thumb' or heuristics by which we can operate efficiently and effectively. These assumptions range from the definitions of things, to the methods we adopt, to the specialised vocabulary that we use to convey our thinking and the models of the world that we employ.

The definition of client and innovation

At first glance, this would seem to be the most obvious piece of shared knowledge to which we could all agree. As it is at the root of the study of clients driving innovation, it must be central to our understanding of the topic. However, it is not always that easy.

Definition of the client

The Penguin English Dictionary (2002) gives the following general definition:

> Somebody who receives the advice or services of a professional person or organisation; a customer.

It then goes on to give the special meanings related to medicine and computing, for example. This seems to be typical of all dictionaries and it provides a useful starting point for a definition for the construction client.

At a meeting of Task Group 58 of the International Council for Research and Innovation in Building and Construction (CIB) in Helsinki in 2005, the following definition was proposed:

> A client is a person or organisation, who at a particular point in time, has the power to initiate and commission design and construction activity with the intention of improving the performance of an organisation's social or business objectives.

This definition tries to take into account that there may be different clients at different points in time who have the role of commissioning construction. It is doubtful whether the *intention* is necessary as presumably all clients undertake such a commission to provide for improvement but, nevertheless, it reinforces the positive nature of the clients' role. Another issue is that one person may well be a client to one organisation and at the same time be a contractor to another, sometimes for the same piece of work. The supply chain in construction often has many such relationships. However, in this book we are focusing on the major clients who initiate and commission the whole building.

In discussion at international symposia, both in the UK and abroad, it has been stated that the definition above is largely Anglo Saxon and that in other countries the word 'client' might have a slightly different meaning (see Chapter 7 for a fuller discussion). For the purposes of this volume, the focus will be on the organisation or person who is the prime initiator of construction activity.

Definition of innovation

Innovation, on the other hand, is now well documented in many books, as the subject has become an active area in management research in recent years. Dictionaries have various insights of which the following are a few examples:

- To make changes by introducing something new (*New Webster's Dictionary*, 1992).
- Any action that occurs spontaneously in a new situation rather than as a result of trial and error learning (*Chambers Dictionary of Science and Technology*, 1992).
- Bring in new methods, ideas, make changes (*The Reader's Digest Oxford Complete Word Finder*, 1993).
- To begin or introduce something new, be inventive (*The Reader's Digest Universal Dictionary*, 1987).
- To make changes by introducing something new, e.g. new practices or ideas (*The Penguin English Dictionary*, 2002).

It is clear from the above that it is the newness of something in an existing situation that determines whether something can be defined as innovation. The new item may be new to that situation but not necessarily to all situations. For example, a new information technology system may be new to a firm or system but it might well have been applied by another firm and in a different context. Innovation is not necessarily invention although invention will nearly always be innovative. Clients are not, therefore, expected to invent something new but rather to introduce something new possibly from another domain. Those construction clients who have come from other industries and bring to construction new ideas from their previous industrial experience are, therefore, being innovative whilst not being inventive.

Many of the texts on innovation explore this matter in greater depth (Chesborough, 2003; Drucker, 1985) and construction has been addressed by several authors including Miozzo and Dewick (2004); Dodgson *et al.*, (2005); and Brown *et al.* (2005, 2006).

It is clear, therefore, that there is a consensus around the person commissioning construction introducing something new to the process or practice or physical artefact that was not previously seen in that situation. This provides the context for this book.

To what degree can the client innovate?

As with most aspects of decision making there is a spectrum of potential input from, in this case, the client. At one end of the spectrum, a client has the opportunity to impede innovation. This happens quite often when the client does not wish to take what he or she perceives is the extra risk of introducing something new into a well-established practice. At the other end, the client may insist that innovation takes place because the current situation is untenable. To continue might endanger life or risk contravening a regulation or legal requirement. Between these two there are a variety of possible interventions.

Figure 1 demonstrates the changing spectrum of how clients could potentially respond to innovation and their willingness or reluctance to drive the innovation process. From left to right the possibilities are as follows:

- *Impede.* Some clients will be 'risk averse' to innovation and will not want to be used (as they see it) as experiments in the construction process. For them the comfort of knowing that traditional processes are being followed with a team that is experienced in such matters is the key issue. Anything else may result in costs to them. Only when others have shown that the innovative system works will they be

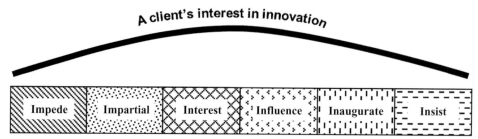

Figure 1   The spectrum of a client's attitude to driving innovation.

willing to adopt the same system for themselves. This is quite a familiar attitude particularly in some government circles where public accountability is an issue.

- *Impartial.* Here the client is not driving the innovation but is willing to listen to those he/she respects if they feel that a new approach will yield advantage. There is no sense of driving the innovative process but merely one of response to suggestion. This can often happen where the client is inexperienced and places more faith and trust in his design/construction team to produce the product or service required.
- *Interest.* There are some clients who are naturally inquisitive about new methods or new technologies but may not have the experience to drive innovation. They show interest in new methods where they see there may be advantage and by showing interest they encourage an environment of improvement not necessarily weighed down by the methods of the past. In addition, they listen to their advisers and respond to their suggestions in a positive manner. Those clients who sometimes come from other industrial/commercial sectors that are naturally more innovative can often be in this category. They do not have the knowledge to drive innovation but they do want to see the best processes and products used for their benefit.
- *Influence.* Although most clients do not build on a regular basis, some do have a stream of work that allows them to understand and encourage new processes or improved performance through their own experience. They can observe the process from a distance and often this allows them to suggest new ways forward. Often these clients have an in-house team of professionals acting on their behalf and they can influence the external team to try new options. In these cases, they do not inaugurate the new method but they have a major impact on the way the external team behaves.
- *Inaugurate.* This is an approach that is a natural extension from influence and it occurs where the client is much more confident about his/her knowledge of the process. This can often happen where an in-house team is building on a regular basis, often with similar types of buildings, and they have the power to adapt and change the design and approach to the total procurement process. Many of the national/international retail and hotel/restaurant chains fall into this category. In addition, it may be that time is of the essence in these developments and the cost risk in construction is not as great as the potential for missing a retail opportunity. If the innovation can speed things up and the revenue can flow earlier, then the saving can often outweigh any additional risk.
- *Insist.* There exist a small but very influential group of clients for whom innovation is the key to their status and success and who genuinely adore the chance to be different to others particularly in the properties that they design. In this category come the clients of large cultural buildings such as opera houses, museums, art

galleries, etc., where the design is part of the whole marketing and enjoyment of the cultural experience. They appoint architects who do not conform to tradition and who produce exciting 'free form' designs with innovative methods and materials. These are the innovations that are given so much attention and that produce the 'shock of the new'. Of course, there will be others who also have agendas to innovate, e.g. for the production of sustainable construction, and these too will demand a high level of new thinking and application.

The above outline, the six 'I's of clients' attitudes to innovation, provides a framework for understanding how clients might react to innovative suggestions. This is a starting point that now needs to be developed in order for a foundation of knowledge to be established for future research to build a thorough understanding of the subject.

This book attempts to provide some clues as to how this foundation might develop. It does not pretend to be a comprehensive volume of chapters that deal with all aspects of the subject. It does not try to establish a thorough and rigorous exposé of the topic as this will follow later. It does not try and encapsulate the thinking of clients in all sectors of the industry as this would be a massive tome and again it is likely that much more research will be needed to even touch on one industry sector.

What it does do is provide insights into what is required to build a base of knowledge from different perspectives. It is exploratory rather than authoritative as this is a subject that is just opening up in the construction sector. It includes work by those involved in government policy. It takes the work of academics who have explored the theory behind the concept of clients driving innovation and it uses case studies to show where successful innovations have taken place that are client driven and looks for the reasons for this success. The reader can select which chapters are most relevant to him or her.

The structure of the book

In order to lead the reader through the subject, the book has been arranged in three main sections:

- Part I (the context for innovation) examines the context for clients driving innovation. This includes a commentary on the theory supporting the idea and the environment in which clients can undertake this role. These issues are so important in building knowledge and placing that knowledge within a framework that allows further development. It is, of course, very broad and in many cases the knowledge is generic. It provides debate on issues that are still open to challenge and it suggests ways forward for further examination. It suggests a taxonomy that could allow a classification for clients, which could allow a different perspective on the approach to be taken for each, and it explores the tools and barriers to the implementation of innovation in practice. These are all important issues when trying to understand the topic and the methods and approaches that are likely to be most successful.
- Part II (the innovation process) is concerned with the innovation process and addresses the important issue of how interventions in the processes of design and construction can yield substantial innovation. If we do not understand when and where to intervene, we will not be able to be effective in improving the clients' role. Case studies are used to illustrate some points from real life projects and research.

- Part III (moving ideas into practice) concerns itself with how the ideas for innovation can be pushed through into practice. This includes some authors who have been involved in policy, some who have had to battle with government, some who have experienced real innovation in the private sector for substantial gain.

The editors are indebted to all those who have contributed and for their willingness to enter into debate and spend the time to put their views down on paper. No attempt has been made to control these views and, in fact, diversity of thinking has been encouraged to provoke new insights into the area. It really is a start to the exploration and there is the potential for this subject to grow in stature for many years to come. One area where more work could be done is the psychology of clients and their behaviour patterns. This is, of course, not just restricted to single clients but also organisations and their corporate culture. It is a rich area for exploration and we would encourage young researchers to enter this arena.

The aim is to improve the construction industry for all its stakeholders whether it is the client or the multitude of participants in the design/construction process. The benefit should also be seen by the public at large, who often have to enjoy or endure the results of the process for many years to come. Indeed, it could be argued that it is the user and the public who are the ultimate 'client' and it is they who should be central to the thinking in this important area. To all who read and all who follow we wish you the very best of luck!

References

Brown, K., Hampson, K. and Brandon, P. (eds) (2005) *Clients Driving Construction Innovation: Mapping the Terrain*. Brisbane, Cooperative Research Centre for Construction Innovation.

Brown, K., Hampson, K. and Brandon, P. (eds) (2006) *Clients Driving Construction Innovation: Moving Ideas into Practice*. Brisbane, Cooperative Research Centre for Construction Innovation.

Chambers Dictionary of Science and Technology (1992). California, Academic Press.

Chesborough, H.W. (2003) *Open Innovation – The New Imperative for Creating and Profiting from Technology*. Boston, Harvard Business School Publishing Corporation.

Dodgson, M., Gann, D. and Salter. A. (2005) *Think, Play, Do. Technology, Innovation and Organization*. Oxford, Oxford University Press.

Drucker, P.F. (1985) *Innovation and Entrepreneurship*. Oxford, Butterworth-Heinemann.

Egan, J. (1998) *Rethinking Construction: Report of the Construction Task Force to the Deputy Prime Minister, John Prescott, on the Scope for Improving the Quality and Efficiency of UK Construction*. Department of Trade and Industry, HMSO, London.

Miozzo, M. and Dewick, P. (2004) *Innovation in Construction: A European Analysis*. Cheltenham, Edward Elgar.

New Webster's Dictionary (1992). Baltimore, Ottenheimer.

The Penguin English Dictionary (2002). London, Penguin.

The Reader's Digest Oxford Complete Word Finder (1993). Oxford, The Reader's Digest Association.

The Reader's Digest Universal Dictionary (1987). New York, The Reader's Digest Association

Woudhuysen, J. and Abley, I. (2004) *Why Is Construction So Backward?* Chichester, Wiley-Academy.

Part 1
The context for innovation

1 A global agenda for revaluing construction: the client's role

Peter Barrett

1.1. Introduction

Construction is often seen as an embattled industry. The repeated critique of numerous reports questions the ability of the construction industry to innovate and manage change to improve its practices. In response to this, a priority theme for re-engineering construction was initially settled upon within the International Council for Research and Innovation in Building and Construction (CIB) in 1997 at the CIB board meeting in South Africa. However, it did not really get off the ground until 2001 when work was put in hand to create a strategy for the development of the theme. This was carried out by Courtney and Winch (2002) and led to a re-orientation around the notion of 'revaluing construction' and a stream of activities that are summarised in Barrett (2007). This latter work involved five workshops, each in a different country, namely Australia, Canada, Singapore, the UK and the US. A postal questionnaire survey (Lee and Barrett, 2006) to the five countries also provides important underpinning.

The main thrust of results of this work on revaluing construction is encapsulated in seven areas for change[1]. Each of these is not in itself extraordinary, but when dynamically linked these have the potential to fundamentally change construction for the good of those involved, their customers and for society. The 'infinity' model given in Figure 1.1 outlines these seven areas and suggests how they must be connected to make progress.

1.2. Infinite options!

The following sections explore the seven areas of the infinity model for revaluing construction with particular reference to the client's role.

1.2.1. Holistic idea of construction

There are starkly different ways of conceiving of construction. The standard industrial classification (SIC) system is the basis of the normal economic perspective and places construction within F45 (construction), a cluster of activities that includes site preparation, building of completed constructions or parts thereof, civil engineering, building installation, building completion, and renting of construction or demolition equipment with operator. The focus is entirely on the physical construction activities. As a consequence, as Winch (2003) points out, 'the bundling of "construction" goods and services used for the SIC is systematically different from that in all other sectors' (p. 652). It

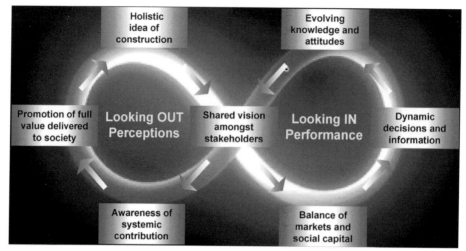

Figure 1.1   Global agenda for revaluing construction.

draws a line between these and intimately linked, value adding activities, such as even the parallel work of architectural and technical consultancies. Crucially, downstream activities such as real estate activities and facilities management concerned with the use phase of buildings are also left out (Ruddock and Wharton, 2004).

An alternative stance starts with the proposition that construction is a change agent for the creation, development, maintenance and operation of the built environment so that it supports the quality-of-life and competitiveness requirements of society. That is, 'construction is a means to a means to an end' (Barrett, 2003). This makes a broader conception of construction entirely logical so that its full contribution to society can be understood. This type of thinking is central to the work of Jean Carassus of CSTB in Paris, who with a group of international colleagues developed and populated a shared economic framework within Task Group 31 of the CIB (Carassus, 2004).

The framework takes the full building life cycle of new construction, management of the service provided by the built environment and demolition. Included in the middle phase are maintenance, major repairs and refurbishment. Taking a vertical view of the framework (from top to bottom) leads from the stream of activities required to create and sustain the built environment, to the panoply of actors or stakeholders with varying degrees of involvement. This ranges from real estate agents and property and facilities managers with an on-going involvement with the asset to those with a short-lived involvement via projects, such as developers, project managers, architects and contractors. Underpinning their activities are the associated contributions of manufacturers and distributors. Lastly, contextualising the whole sector are the institutional actors at various geographical levels together with professional representative organisations and user associations themselves. These together infuse the sector's norms, regulations and expectations.

The thrust of Carassus *et al.* is to shift thinking from an 'industry' focus on simply building buildings to a 'construction sector system' approach with the emphasis on producing and managing the *services* rendered by these structures throughout their life cycle to support an efficient and sustainable economy. Positively, in the UK the Strategic Research Agenda published by the recently created National Platform for the Built Environment (2006) has already drawn from this aspect of the revaluing construction work, doubling its estimate of the size of construction from 10% to 20% of gross domestic product and focusing broadly on the built environment, including the use phase.

Within this service-orientated view, clients, users and facilities managers hold a vital position in driving for the type of built environment they want and need and so defining the capabilities of the construction industry required to bring this about. Of course, clients vary considerably and so it should be expected that complex, sometimes contradictory, views will emerge, but this does not undermine the centrality they must take in any analysis.

1.2.2. Shared vision amongst stakeholders

The broader conception of construction set out above can inform the creation of a shared vision amongst key stakeholders for the maximising of value across the whole life cycle of constructed artefacts. It is suggested that a re-valued industry will maximise the initial creation of *potential* value in a particular building/project through pre-design and design activities, its *delivery* through construction, *realisation* in use and *synergies* with other developments at an urban level.

To inform the wide-ranging scope of this conception of construction, it is clear that a broad constituency of stakeholders needs to be engaged to develop and sustain a vision for construction. Moving beyond the construction industry itself defining a strategy, it is clear that without the manufacturers and suppliers involved many opportunities for improved efficacy and efficiency would not surface, without active involvement from clients and users effectiveness is likely to be inadequately addressed, and without societal representation in terms of planners and representative groups the ethicality of proposals is likely to be underdeveloped.

Lastly, without some design input the competing demands of the multitudinous and diverse stakeholders are unlikely to resolved into elegant solutions. On this last point, two aspects need clarification. First, the term 'elegant' is meant conceptually and practically in that the solutions involve sufficient complexity, but are as simple as they can be. Second, 'designers' here are not building designers, but rather 'leaders as designers' (Senge, 1990, pp. 341–345) at various levels who understand the complexities of the 'system' and put together the ideas and the infrastructure in ways that enable the industry to perform to its full potential. At the top level come shared visions held jointly by key national, leading, stakeholders; here the client influence is key.

Bringing together the relevant groups and engendering productive co-working is a complex political process. The form of the Danish clients' association raises an issue that moves the discussion onto a different level. In the UK, the creation of CRISP (Construction Research and Innovation and Strategy Panel) was seen as the solution with a single forum within which the main players in the industry could generate a strategic view. The Danish clients' forum is strongly independent, and linking this point to the wide range of stakeholders with something to offer, indicated earlier in this section, the notion of a dialectic among various stakeholder voices seems a looser, but arguably a more dynamic and richer way is to imagine creating a consensus around a meaningful shared vision. Ideally, the scope of such an arrangement would have a strong role for clients and service providers, but with a clear axis with government policy leadership for the necessary motivation and action to occur. It is common to say that the industry is unduly fragmented, but it is common too for the government interests in construction, such as planning, housing, construction itself, etc., to be located in different ministries and to have diverse remits and concerns. So, it is reasonable to suggest that the leadership provided by governments at a policy level often needs to be harnessed collectively so that a clear dialogue can be held with the other actors. The third axis is to education and research, primarily through universities. There is available a wealth of relevant knowledge about practices worldwide, robust conceptual models, experience in other sectors, and educational solutions for long-term change. This dimension completes a balanced set of groups that represent different perspectives and that can through a debate, not devoid of tensions, fashion a robust, shared vision.

Creating a vision is one thing, but VTT (2005) reviewed 16 construction industry strategies from around the world and found that 'implementation is barely covered' (p. 10). For a sustainable vision to robustly underpin medium-term transformation, the workshops highlighted the importance of emphasising equitable returns to all stakeholders, as well as the sheer maximisation of the collective value created. Being fully inclusive is a good first step though. If parties are left out, as has often been the case in the past, then they will not be harnessed as advocates. It is interesting to note in this context that an international postal questionnaire survey about innovation in construction (Lee and Barrett, 2006) highlighted that institutional bodies 'protecting their own interests' were the second highest rated inhibitors of the adoption of new practices, but the highest were clients 'protecting their own interests'. Thus it is crucial that a client voice is actively encouraged and engaged with.

1.2.3. Balance of markets and social capital

Clients have a key role in how the market for construction work operates. Cut-throat competition, unscrupulous low bidders, lack of trust and poor risk management are very commonly quoted characteristics of construction. These are implicitly reinforced by standard contracts, custom and practice and the sheer confrontational nature of the industry. Do the benefits accrued justify the direct and opportunity costs involved?

The generic question about the 'pay-offs' associated with the pursuit of short-term personal gain is addressed in the 'prisoner's dilemma' experiments with their roots in game theory (Kay, 1993, pp. 35–49). The main features of the successful cooperative strategies in iterated games are as follows: the participants begin by expecting the other player to cooperate, not that he or she will cheat; they respond to bad behaviour and punish it, but not too severely; and they are forgiving. Interestingly at the broader societal level, Sacks (2002) makes a parallel argument for valuing relationships borne of on-going interactions (pp. 149–160). These, he argues, will be created at a local level within groups of people that share a bond that goes beyond individual transactions and could be called 'social capital' or the level of trust in a society. Sacks (2002) takes the argument beyond the benefits for contractual dealings and distinguishes these from what he terms 'covenantal' relationships. These are often evident in family life, but in broader life are sustained by loyalty, responsibility, fairness, compassion – professionalism at its best. Without these trust-based, reciprocal relationships and the institutions that support them, 'markets and states begin to fray. . . social life itself loses grace and civility'.

So, how does this relate to construction? The commercial environment for construction has been typified as 'competition is good; more competition is better' (Ang, 2004). This resonates with Hendriks' (2004) phrase 'the poison is the dose', that is, something that in moderation can be beneficial, in too great a quantity can be fatal. Given these aggressive conditions, it is hardly surprising that 'lock in' is felt by the individual players. Who can afford to move first?

Well maybe the answer is the clients, but why should they? Some interesting work has been carried out by Zaghloul and Hartman (2003) in which the effect of 'disclaimer clauses' in contracts between clients and contractors was studied. These clauses are extensively used in traditional and new partnering style contracts to shift risks to the contractor on issues such as delay and uncertainty of work conditions. The study was based on a survey of more than 300 industrialists in North America. It revealed that contractors assess the five most commonly used of these clauses such that a premium of between 8% and 20% is added to their price in a seller's market. This is a very clear and tangible measure of the significant cost of risk within construction relationships. Interestingly, the authors found that low levels of trust were typical, but that where a high level of trust did exist the impact on the premiums was profound. The premiums in these circumstances were 'very low' because the perception of the shift in risk was from around 4.4 to only 2.2 (on a five-point scale). The benefits of a high level of trust were found to be facilitated if the parties had previous experience of working together such that they felt the other party displayed competence, integrity and, more generally, a good reputation within the industry. In ideal, but rare, circumstances the risks were intelligently and equitably distributed without recourse to blanket clauses. So, for clients the value of building trust is clear and the route forward through some form of on-going relationship has been highlighted as a significant practical variable.

The consensus from the workshops in five countries was that the key objective was to procure flexibly, optimally and appropriately. This is a complicated way of saying that just using the 'normal' approach, because one always does, is not good enough. Procurers need to assess each situation and be flexible enough to establish the optimal approach that is appropriate for the given situation. This will be conditioned by many things, but is likely to take into account the value of long-term relationships. This can be realised through pre-qualification using a range of key performance indicators, which can include the track record or capacity of the players to work well as a team. Selection based on broader criteria than simply price is becoming quite prevalent, albeit in two-stage tendering price is likely to dominate in the second stage.

Partnering/alliancing arrangements are a move in this direction, but a move to less contractual, more fundamental shared social norms, rooted in professionalism and codes of ethics, is an area that deserves more attention. Practical outcomes are likely to be the selection of project participants based on all-round performance, not simply lowest price, and an equitable allocation and management of risk. For this to happen, accessible and appropriate norms, metrics and contracts will be needed. However, it should be remembered that it is quite possible to move too far in the direction of cooperation as the recent problems of collusion, or 'horizontal integration', in the Netherlands (PSIBouw, 2004) have illustrated. This issue has also been picked up at the generic management literature in terms of the 'dark side of close relationships' (Anderson and Jap, 2005); thus, the emphasis in the title of the section on 'balance'. It is, however, only really the clients who have the power to shift this balance.

1.2.4. Dynamic decisions and information

In tandem with the above market considerations, there is a need for dynamic decisions and information throughout the building life cycle. Currently, there are huge gaps in the process through which information and understanding are lost at very great cost. Decision-making is undermined and this is compounded by the combative culture of the industry. However, it was clear from the revaluing construction workshops that, without a conducive procurement context, those working at the project level are almost certain to adopt defensive routines that minimise their risk, but close down opportunities to maximise the joint value created. However, once these conditions have been realised in whole or in part, significant progress can be sought through increased integration using ICTs and the adoption of an appropriate team value management (VM) approach. The common theme to these two foci is seamless integration, focused on, respectively, information/technology and people/decisions.

In the area of information and technology, a report by the National Institute of Standards and Technology (NIST) (Gallaher *et al.*, 2004) dramatically highlights the potential for improvement by assessing the costs of inadequate interoperability in the US capital facilities industry. The headline figure revealed is $15.8 billion 'lost' every year, or 1–2% of industry revenue. The report covers only commercial, institutional and industrial facilities, but the analysis extends for the whole life cycle, from planning and design, through construction to operations, maintenance and decommissioning. Interestingly, the results indicate that two-thirds of the 'unnecessary' costs are met by owners and operators of buildings, whereas architects and engineers are only associated with about a tenth of this impact. The balance of around a quarter of the costs is met by the industry in terms of contractors, specialists and suppliers. The clear implication raised in the report is that 'interoperability costs during the O+M phase [result] as a failure to manage activities upstream in the design and construction process' (Gallaher *et al.*, 2004, p. 25). The figures of course also indicate that motivation to move on the issue is skewed in that the greatest potential to change things is in the area where the least cost is experienced. This is an open invitation to sophisticated clients to demand better use of data models so that the downstream benefit can be released.

Briefing needs content as well as a data structure and Horgen *et al.* (1999) describe a rich, team-based approach to briefing that highlights various challenges, including keeping existing organisational processes 'unfrozen' long enough to allow the optimal built solution to emerge. Luck (2002) has worked on a close analysis of the language used during interactions with users in briefing and concludes that 'design knowledge cannot be completely represented in a prepositional, non-contextual form' (p. 16). Barrett and Stanley (1999) argue for the necessity of allowing time for trust to develop so that there is time for co-learning through linked processes of disclosure and feedback. So, briefing has to be a dialogue. How extended and open-ended this is will depend on the novelty of what is intended, but also the knowledge, experience and degree of mutual understanding of the main participants. It is, of course, essential that the above is not used as an excuse for procrastination. So defined, the briefing process articulates strongly with the building production phase, which is the next area for discussion.

The benefit of VM in the production phase was often mentioned at the workshops in five countries. However, it is necessary for VM to impact during the whole of the construction process and this is reinforced by Neff (1998) in his advocacy of the STEPS approach, where the usual political, economic, social and technological factors are supplemented by a demand to find 'synergies'. Neff's argument is illustrated by a case study of a complex sub-service project and gives a recommendation that the STEPS factors need to be reviewed and risks managed progressively with clear mitigation measures and controls. It is in these latter areas that the potential for synergies becomes evident, especially as the practical complexities and constraints mount through the project period. In general, little has been said about continuity beyond initial VM efforts fairly early in the project, although many from the design side of the industry see VM workshops as horrendous, destructive events. The challenge must be for VM to become a more natural part of the thinking and practices of those involved, rather than a confrontational, occasional event. Neff (1998) stresses that for this type of joint problem solving to work, good will and trust are demanded.

Clients, by creating the conditions and expectations for regularly returning to refresh the shared understanding of their aspirations, and steering the project against these, can orientate construction activities much more powerfully towards their ends. By insisting on coherent data structures, the information and knowledge so created can be captured for the benefit of the use phase.

1.2.5. Evolving knowledge and attitudes

The workshops in five countries uncovered profound concerns about the workforce available for construction; for example, a trade skill and, looming, age gap; a lack of design and construction management skills; and a lack of top management capacity. In terms of clients there were particular concerns about the industry's failure to understand them and their lack of ability to clearly express their requirements to the industry. The notion of 'educating "about" clients' covers both ends of the relationship. Clients need educational opportunities to be better clients and a client orientation should infuse the development of all construction personnel.

To support the linkage between universities and industry, there should be joint forum/s so that a dialogue can be maintained and a mutual understanding of the participants' distinctive roles allowed to develop. However, there are very few powerful drivers behind the educational aspect of the picture, but action here is critical to the success of the overall revaluing construction effort. People are the industries' most valuable raw material. Dainty *et al.* (2004) go as far as to call the situation in the UK 'the construction labour market crisis', which demands collective action. Knowledgeable clients have a key role to play, as does government, as both can use their influence to reinforce the involvement of educational providers in key forums to stress the importance of understanding clients and involve clients themselves.

1.2.6. Awareness of systemic contribution

Clients drive demand for construction and as such aim to enjoy benefits, but also carry some responsibilities. Of course, construction results in physical artefacts and these are typically created in a regulatory context that seeks to ensure minimum standards on aspects of public concern, such as structural stability, health and safety, energy conservation and accessibility. However, beyond these minimum standards there is a potential to create spaces that to varying degrees enhance the quality of life and competitiveness of those using the spaces created in terms of physical, functional and psychological effects.

These can be difficult to measure, but there are general studies that provide pointers, for example, in the area of higher education where CABE (2005) provide feedback from case studies of five UK universities and highlight the important design dimensions for the recruitment, retention and performance of staff and students. Interesting differences in views are revealed as staff appear to be more attracted by buildings with a 'feeling of space . . . aesthetic appeal', whereas students stress more functional aspects: 'modern design . . . quality of facilities" (pp. 46–47). In addition to differing views, there can be conflicting criteria, exemplified in Heerwagen *et al.*'s (2004) paper on creating collaborative learning environments that concludes that the central dilemma remains that 'spaces designed to increase awareness and interaction also increase the potential for interruptions and distractions' (p. 525).

However, a rare and good example of work that does cut a way through some of these issues is provided by Zeisel *et al.* (2003) in a paper that identifies environment–behaviour links between the design features of care facilities for Alzheimer's patients in the US and behavioural health outcomes, such as anxiety, withdrawal, depression and aggression. In general, these dimensions of behavioural health problems are seen to recede in the face of design features such as providing for privacy, variability in social spaces, careful (camouflaged) exit design and creating a residential rather than 'institutional' feeling environment. There are distinctions to be made; for example, verbal rather than physical aggression appears to correlate with the environmental design. This type of work provides very valuable insights and although it was for a particularly vulnerable group similar, less stark, issues would doubtless translate, say, to the office environment. By supporting such studies clients could create the potential for evidence-based design decisions to meet their needs.

Widening the scope beyond specific organisations, clients have opportunities and responsibilities as built artefacts make major impacts on societies. Issues such as regeneration can transform lives (and property values) when individual developments add up to attractive spaces/areas for everyone. Some are beginning to coin the phrase 'Urban FM' to highlight the need to actively manage some of these spaces too, to gain optimum impact. Clients with a focus on property as an investment will be central to the response to wildly varying demographic trends. For example, population growth from 2000 to 2050 is predicted to swing from over +40% in the US to −30% in Russia. Within these figures there are major shifts in the age profile with, in the developed world, an increase in the percentage of the population aged 65 and over from around 8% in 1950 to around 20% in 2020. An ageing population implies different household sizes and adaptations to support the elderly living independently in their own homes. If this is not addressed there are major implications for their quality of life and a large bill for formal health care provision (Lansley *et al.*, 2004). The implications for the built environment are profound, from creating housing of at least the minimum standard for the world's poor, to adapting the built environment to huge quantitative and qualitative population changes.

Taking the still broader perspective of the planet and environmental issues, the built environment is responsible for consuming 40% of all the raw materials taken out of the Earth's crust, by weight, 40% of waste streams, and 40% of greenhouse gas emissions (Milford, 2004). These are huge impacts and, of course, there are tremendous opportunities to reduce these impacts by more efficient processes, recycling materials and re-use. There are major moves to increase life cycle and embodied energy performance and these should lead to reduced environmental impacts. In almost all cases clients can take a lead by demanding and valuing good practice in these areas; without this, lowest cost will probably drive out change until the situation is so severe that legislation is used to get action.

Construction creates physical artefacts and often the most prominent dimensions of this are the mess and disruption caused during construction and the cost of the works. However, a pervasive stream of soft and hard benefits also flows over the longer term and needs to be better accounted for so that the return on the initial investment can be made more clearly visible. In this connection key gaps that need attention are the creation of a comprehensive framework for these value streams and the clear identification of appropriate ownership amongst stakeholders. This is an area where clients could lead and in so doing shift their investment decision-making onto an improved evidence-based footing.

1.2.7. Promotion of full value delivered to society

There have been many critical reviews of construction in just the UK (Murray and Langford, 2002) and books such as Woudhuysen and Abley's (2004) 'Why Is Construction So Backward?' jump on the critical bandwagon. This is despite evidence to the contrary; for example, a survey by the NRC of major construction clients in Canada (Manseau, 2003), which found 94% were satisfied or very satisfied. Similarly, a survey of 200 clients in the UK (Barrett and Ruddock, 2000) indicated a robust performance, but with room for improvement on the service dimension.

A core debilitating assumption is that the industry is unhelpfully fragmented. However, it is suggested here that the industry is actually highly differentiated to meet the diverse and complex demands placed upon it, and that at a project level the integration effort is generally kept in clear focus. It could be said that 'single loop learning', that is pragmatic problem solving to 'do things right', on the ground is alive and well. The same cannot be said for longer term company-based innovation or the policy framework within which it is placed. Here 'double loop learning' is severely limited by the turbulence of the industry's workload and the limited resources of small- and medium-sized enterprises so that progressively moving towards 'doing the right things' is hard to sustain. If clients can create longer term relationships this situation could be improved. If clients more vociferously backed the industry when it does a good job that too would make a big impact that would have wide-ranging benefits, including an improved self-image of those in construction and greater attraction for high-quality new recruits.

1.3. Conclusions

Clients can have a significant impact, both in relation to their own projects and, for some, as drivers to policy reform shaping the context within which others work. In this latter area, the government has a responsibility as the major client in most countries to use this influence to good effect. But any client can engage with the policy debate and ensure that it reflects the needs and aspirations of users, owners and investors. This can happen at a local or national level and will be particularised to the specifics of the sub-sector in question.

To address how clients can act to achieve significant change in their own projects, examples of exemplary practice will be drawn upon that emerged from the workshops (Barrett and Barrett, 2006). It was clear from a cross-case analysis that improved collaboration or innovation alone could have significant impacts reflected in phrases such as 'all involved gained' and the 'system worked incredibly well'. The incidence of these isolated themes was, however, the exception and it was much more common for the examples to display a combination of collaboration, or collaboration and creativity, *driven* by severe constraints. This seemed to reflect situations where those involved are under great pressure, but through collaboration, sometimes with creativity added in, the challenge was successfully met. Sentiments were expressed, such as 'the whole team simply pulled together', 'the feeling of working in such a way was great', 'initiative and energy of colourful and talented builders', 'very unusually' and 'discarding the traditional approach'. These convey the feeling of strong social bonds flourishing once the stifling limitations of 'normal business' were relaxed in the face of extreme demands.

It is interesting to note that in several cases of constraints, a way forward was fashioned by actually releasing a significant limitation, such as taking the second lowest bid on a project that was very high profiled, guaranteeing a minimum price or explicitly taking time out to fully work through the briefing/design issues. Further, it was quite common for projects to have explicit community (intra or extra to the organisation) benefits. So, 're-valued' construction can be typified by projects where significant constraints drive those involved to collaborate strongly, spurring the team to innovative responses that not only triumph against the demands of the project itself, but also impact positively on the community around.

A major aspect within all of this is the role of constraints and the interesting question, in the project arena, as to when is a constraint a positive factor and not a restraining force? The answer would seem to be when it is stated explicitly, clearly and early on. Further, when it is demanding enough to define the project, prioritise and re-orientate behaviour around a super-ordinate goal and provide a clear measure of success. This is aided by a situation where other (less important) parameters are dealt with flexibly so that appropriate collaboration and creativity, both technical and organisational, are facilitated to meet the challenge. In this way, the constraint has provided a clear space to work within and this certainty can clearly be stimulating provided sufficient flexibility with the remaining resources is available. This contrasts with most restraining forces that sap energy and hold back initiative and hamper creativity. Another pertinent aspect is the nature of the constraint itself. Far from being arbitrary, participants clearly understood and accepted the rationale for these restraints, whether social, time related or environmental. Success with such a task is then explicit and the pride of the participants in these projects was clearly evident as they told their stories.

Before ending, it is important to distinguish the differing nature of the two halves of the infinity diagram. The right-hand 'looking in' side stresses actions to enhance the performance of the industry and as such to deal with complex issues for which simple, uni-dimensional solutions are not available. Thus, the signature of this half is 'appropriateness' or balance, between various factors. What this chapter aims to do is pinpoint the main aspects around which the issues in each area appear to rotate. However, the argument is categorically *not* that long-term relationships should replace contracts, that all information should be kept and recycled, or that education and re-search alone can solve the problems of the industry. Centrally, it is also *not* arguing for a single monolithic vision for the industry, but rather a vibrant dialogue with energised partners, amongst which clients are a very important player. Some level of coherence and focus is, however, desirable and the schema set out in this paper is a suggested landscape within which, for example, the conflicting views evident in the special volume of papers on revaluing construction (Courtney and Winch, 2003) can be fruitfully pursued and practical actions identified.

The 'looking out' half of the diagram is different in emphasis and broadly speaking is underpinned by an argument for looking broadly beyond existing categorisations, mindsets and images. This is evident in the 'holistic' area where existing economic conventions are problematic, in the 'systemic' area where the accounting of contributions from construction is lacking, and in the 'promotion' area where the industry is significantly undervalued, by others and by itself. In particular, the theory and practice of accounting for the value delivered by construction in its fullest sense is a rich vein to be pursued, for example, as has Saxon (2005).

The notion of 'revaluing construction' has, through its very lack of definition, allowed a wide range of ideas and issues to be fruitfully connected. Central to the emerging view has been a conclusion that revaluing construction concerns 'the maximisation of the value jointly created by the stakeholders to construction and the equitable distribution of the resulting rewards' (Barrett, 2005), or in short 'creating value for all'. Provided that performance improvements broadly match positive changes in perception, then a virtuous cycle can be expected to operate. However, a concerted effort over a number of years or even decades will be needed to achieve significant and enduring change and it will be essential that clients are intimately engaged.

Notes

[1] For a fuller discussion of the seven areas for change, please see Barrett (2008).

References

Anderson, E. and Jap, S.D. (2005) The dark side of close relationships. *MIT Sloane Management Review* **46** (2): 75–82.

Ang, G.K.I. (2004) Competing revaluing construction paradigms in practice. In: *Revaluing Construction* (ed P.S. Barrett). Oxford, Blackwell Science, pp. 83–104.

Barrett, P.S. (2003) Construction management pull for *n*D CAD. In: *4DCAD and Visualization in Construction* (eds R. Issa, I. Flood and W.J. O'Brien). Heereweg, Swets & Zeitlinger, pp. 261–280.

Barrett, P.S. (2005) *Revaluing Construction: A Global CIB Agenda*. Rotterdam, CIB.

Barrett, P.S. (2007) Revaluing construction: an holistic model. *Building Research & Information* **35** (3): 268–286.

Barrett, P.S. (2008) *Revaluing Construction*. Oxford, UK, Wiley-Blackwell.

Barrett, P.S. and Barrett, L.C. (2006) The 4Cs model of exemplary construction projects. *Engineering, Construction and Architectural Management* **13**: 201–215.

Barrett, P.S. and Ruddock, L. (2000) *Building Surveyors' Customer Survey*. London, RICS.

Barrett, P.S. and Stanley, C. (1999) *Better Construction Briefing.* Oxford, UK, Blackwell Science.

CABE (2005) Measuring the impact of architecture and design on the performance of higher education in institutions. In: *Commission for Architecture and the Built Environment* (CABE). London, CABE.

Carassus, J. (ed.) (2004) *The Construction Sector System Approach: An International Framework.* Rotterdam, CIB.

Courtney, R. and Winch, G.M. (2002) *CIB Strategy for Re-Engineering Construction.* Rotterdam, CIB.

Courtney, R. and Winch, G.M. (2003) Re-engineering construction: the role of research and implementation. *Building Research & Information* **31**: 172–178.

Dainty, A.R.J., Ison, S.G. and Root, D.S. (2004) Bridging the skills gap: a regionally driven strategy for resolving the construction labour market crisis. *Engineering, Construction and Architectural Management* **11**: 275–283.

Gallaher, M.P., O'Connor, A.C., Dettbarn, J.L. and Gilday, L.T. (2004) *Cost Analysis of Inadequate Interoperability in the U.S. Capital Facilities Industry.* Gaithersburg, MD, National Institute of Standards and Technology.

Heerwagen, J.H., Kampschroer, K., Powell, K.M. and Loftness, V. (2004) Collaborative knowledge work environments. *Building Research & Information* **32**: 510–528.

Hendriks, F. (2004) The poison is the dose or how 'more egalitarianism' may work in some places but not in all. *The European Journal of Social Science Research* **17**: 349–361.

Horgen, T.H., Joroff, M.L., Porter, W.L. and Schon, D.A. (1999) *Excellence by Design Transforming Workplace and Work Practice.* New York, John Wiley & Sons.

Kay, J. (1993) *Foundations of Corporate Success.* Oxford, Oxford University Press.

Lansley, P., McCreadie, C., Tinker, A., Flanagan, S., Goodacre, K. and Turner-Smith A. (2004) Adapting the homes of older people: a case study of costs and savings. *Building Research & Information* **32**: 468–483.

Lee, A. and Barrett, P.S. (2006) Value in construction: an international study. *International Journal of Construction Management* **6**: 75–86.

Luck, R. (2002) Dialogue in participatory design. Presented at *Common Ground Conference,* September, University of Brunel, London.

Manseau, A. (2003) *Survey of Major Clients: Improvements and Innovation in Construction Investments.* Toronto, NRC.

Milford, R. (2004) Re-valuing sustainable construction. In: *CIB World Building Congress – Building for the Future.* Toronto, CIB.

Murray, M. and Langford, D. (eds) (2002) *Construction Reports 1944–98.* Oxford, UK, Blackwell Science.

National Platform for the Built Environment (2006) *Research Priorities for the UK Built Environment.* London, Constructing Excellence.

Neff, T.L. (1998) Risk management considerations for complex sub-surface projects. In: *Subsurface Conditions Risk Management for Design and Construction Management Professionals* (ed D.J. Hatem). New York, John Wiley & Sons.

PSIBouw (2004) *Dutch Construction Industry Reform Programme.* www.psibouw.nl. Rotterdam.

Ruddock, L. and Wharton, A. (2004) Revaluing construction, drivers from the urban environment – construction and sustainable development. In: *CIB World Building Congress.* Toronto, CIB.

Sacks, J. (2002) *The Dignity of Difference.* London, Continuum.

Saxon, R. (2005) *BE Valuable: A Guide to Creating Value in the Built Environment.* London, Constructing Excellence.

Senge, P.M. (1990) *The Fifth Discipline the Art & Practice of the Learning Organization.* London, Random House.

VTT (2005). *Pre-Summit Study: Global Synthesis Report.* Helsinki, VTT.

Winch, G.M. (2003) Models of manufacturing and construction processes: the genesis of re-engineering construction. *Building Research & Information* **31**: 107–118.

Woudhuysen, J. and Abley, I. (2004) *Why Is Construction So Backward?.* Chichester, John Wiley & Sons.

Zaghloul, R. and Hartman, F. (2003) Construction contracts: the cost of mistrust. *International Journal of Project Management* **21**: 419–424.

Zeisel, J., Silverstein, N.M., Hyde, J., Levkoff, S., Lawton, P.M. and Holmes, W. (2003) Environmental correlates to behavioral health outcomes in Alzheimer's special care units. *The Gerontologist* **43**: 697–711.

2 Revaluing construction: implications for the construction process

Graham M. Winch

2.1. Introduction

Since 2002, the International Council for Research and Innovation in Building and Construction (CIB) has been developing a theme entitled revaluing construction. Since the publication of an initial report in that year, three conferences on the theme have been organised, and a considerable intellectual activity mobilised. Why has this theme caught the imagination of the research and reflective practitioner community? Arguably, this is because it weaves together a number of the different elements found in various national construction industry reform programmes around the world into a coherent pattern that can then be articulated as a vision for the future of the much-maligned sector. This chapter will briefly set the context of revaluing construction before turning to focus on the implications of the concepts for the construction process.

2.2. From reengineering to revaluing construction

During 2001 and 2002, the CIB funded a research proposal led by a team from the then Manchester Centre for Civil and Construction Engineering, UMIST (University of Manchester Institute of Science and Technology), UK. The brief was to examine the reengineering of the construction process in analogy with the more generic business process reengineering (BPR), which had become popular in the early 1990s. BPR emerged amongst consultants advising on the implementation of IT systems focused on designing customer-orientated processes that eliminated waste (e.g. Hammer, 1990). Davenport (1993, p. 2) defines more broadly process innovation as 'the envisioning of new work strategies, and actual process design activity and the implementation of the change in all its complex technological, human and organisational dimensions'. While BPR in particular, and process innovation more generally, have had some notable successes in improving process performance, they have also met stiff resistance and been associated with large staff reductions in reengineered companies. The advocates of BPR also suggested that an incremental approach was unlikely to achieve the level of change required for continued competitiveness and advocated discontinuous change. In parallel with the development of BPR, researchers examining the phenomenal success of the Japanese car industry in international markets (Womack *et al.*, 1990; Womack and Jones 1996) defined the concept of *lean manufacturing* where a focus on customer needs and the elimination of waste although a more incremental approach are advocated. It was in this context of these process-orientated analyses of production that reform initiatives were launched in a number of national construction industries, frequently inspired by the UK's Egan Report (1998), which took these emerging production concepts and used them to develop an engaging vision of the future for the construction sector.

Arguably the most innovative and most lasting aspect of BPR is the relentless focus on the process of the delivery of goods or services to customers, and it is this process focus that soon emerged in our research as a major issue for those participating in the CIB research process. This consisted of a series of international workshops; an on-line expert survey; an international conference held in Manchester, UK, in February 2003, and the commissioning of a special issue of *Building Research and Information* (Vol. 31, 2, 2003). Since the reception of the UMIST report by CIB, the activity around the development and implementation of its recommendations has been co-ordinated by the University of Salford (http://www.revaluingconstruction.scpm.salford.ac.uk/), and further policy conferences have been held in Rotterdam in 2005 and Copenhagen in 2007. These were complemented by a second series of international workshops co-ordinated from Salford during 2004 (Barrett, 2007).

What emerged from the UMIST research team's enquiries was that in an industry such as construction where the product typically (but not always) underwent considerable adaptation to meet the needs of clients, *product* was as important as *process* in production. In a sector where the market was dominated by proactive clients as opposed to reactive customers, a total process focus was an inappropriate means of addressing client needs. It is notable that the lean production and BPR concepts had emerged in mass services and manufacturing focused on meeting the needs of large numbers of customers whose requirements do not differ greatly, rather than in those sectors focused on meeting the needs of clients whose requirements do differ significantly. The difference can be crystallised around the marketing problem which in mass customer-orientated markets is to disaggregate the market to find those segments which can be more profitably exploited, while the marketing problem in client-orientated markets is to *aggregate* the market to find as many different clients as possible with closely related needs.

From the discussions in the workshops emerged the concept of revaluing construction – coined at a meeting hosted by VTT in Finland by David Hall of the British Airports Authority (BAA) in December 2001. Although there is (as yet!) no Wikipedia definition of revaluing construction, one perspective is that its fundamental contribution is to marry the process-focused concerns of business process redesign with the product-focused concerns around how buildings add value for clients (Spencer and Winch, 2002). Thus its concerns are holistic (Barrett, 2007) with respect to our understanding of the creation of the built environment. It is also concerned to stimulate a reappraisal of the construction sector in its societal context emphasising that it does not have to be 'dirty, demanding and dangerous' but that it is a major source of wealth creation and can provide attractive careers for ambitious young people.

Before moving to analyse the implications of revaluing construction for the construction process, it is worth examining what exactly is meant by the word 'value'. The revaluing construction concept puns two different but related meanings of 'value'. The first is value as an economic concept of worth; the second is value as an ethical concept of consistently held belief (Ramirez, 1999). Our focus in this chapter is on the first of these, but we will inevitably be drawn into the links with the second during the analysis. The current debates within the construction sector – see the useful review by Thompson *et al.* (2003) – typically deploy an economic concept of value that is directly derived from (Smith's 1970, p. 131) 18th century classical distinction between 'value in use' and 'value in exchange'. The argument (e.g. Rouse, 2004) is then developed that exchange values for buildings do not fully reflect their use values to their occupiers and the two are seen as in tension and there is a slippage towards 'economist-bashing' (O'Hare, 1997) in the debate. Arguably, this distinction is a product of seeing the constructed product as an artefact analogously to the belief of the classical economists that only tangible objects can have value (Barber, 1967). The development of neo-classical economics with its emphasis upon price formation through the marginal equalisation of supply and demand made the distinction between exchange and use value irrelevant because *utility* was determined by what the market was prepared to pay to satisfy its wants (Barber 1967).

This concept of utility is central to the investment appraisal approaches, which are mandatory for public sector clients in most countries as exemplified for the UK in the HM Treasury Green Book and form the basis of whole life management techniques (Boussabaine and Kirkham 2004). The net present *value* of an investment is the difference between the time-discounted expected costs and expected benefits associated with that investment. The argument for treating buildings as assets rather than artefacts, therefore is not merely an advocacy of 'good design' where use value should be preferred over exchange value (e.g. Macmillan 2006) but an argument that we need to develop a much deeper understanding of the benefits of good design so as to enhance the returns flowing from the expected benefits to fund any additional costs that might be incurred in achieving them through 'value added' investment (Spencer and Winch 2002, Figure 3.2). thereby providing a net increase in utility. In this perspective, any inability to trade the completed building is a problem in information in that potential users who value the building as highly as the current users have not yet been identified. Deeper and more widely diffused understanding of how buildings add value for clients should, in principle, reduce these information problems.

Neo-classical economics has traditionally focused on the role of markets in co-ordinating the allocation of resources, and tended to ignore how value is created through co-ordinated processes of production (Milgrom and Roberts, 1992). Yet value is created through the activities of firms, and an understanding of how they do this is vital for exploring the implications for the construction process of revaluing construction. Our framework for this exploration here is taken from a development of the research by Clark and Fujimoto (1991) into new product development in the car industry. They identify two aspects of *product integrity*: the internal integrity of the car design as an engineered system, and the *external integrity* of the design as an expression of potential customers' values. This concept was developed for application to constructed products by Winch (2002) and complemented with the concept of *process integrity* defined as the organisation of the process that achieves product integrity.

2.3. Revaluing the construction product

We turn first to a brief review of the issues around revaluing the construction product. The essence of the argument here is that constructed products have typically been perceived as artefacts rather than assets (Spencer and Winch, 2002) where artefacts are perceived as things that have no capacity to create further value for their owners and users. While they might have a resale value as artefacts, they are seen simply as a cost of doing business and hence something to be minimised. Or, where the decision has been taken not to minimise costs, the additional expenditure is seen merely as a choice regarding the level of current consumption rather than a higher level of investment to gain greater returns. Assets on the other hand have the capacity to create further value due to their design. They are an investment rather than a cost, not simply because property values outperform other investment opportunities, but because greater investment can return greater benefits from the exploitation of the asset to provide services valued by users.

This concept of the artefact is embedded in the way we talk about buildings. Our thinking is still strongly influenced by the Vitruvian triangle popularised by Alberti of *utilitas, venustas,* and *firmitas.* While translations of these terms vary, they capture contemporary thinking about design quality and formed the basis for the development of the UK system of design quality indicators (www.dqi.org.uk), which captures and elaborates Vitruvius in the trinity of functionality, impact, and build quality (Prasad, 2004; Gann *et al.,* 2003). However, the definition of the DQI sub-categories makes it clear (Dickson, 2004) that they are all descriptors of the building as an artefact and do not capture the value generated from that artefact. For instance, there are no criteria that attempt to address how the building increases the efficiency and effectiveness of the staff working in it, yet this is far and away the largest cost of a building through its life (Wright *et al.,* 1995). Or, to put the point more broadly, there is little indication of how the design quality indicators are linked to organisational performance in terms of better educated children, more customers walking through the door, swifter service response times, or more stimulating theatrical experiences.

Rouse (2004) has reviewed some of the constraints in addressing the value generated by buildings during the investment appraisal process and hence treating them as artefacts rather than assets, but, arguably, the largest constraint is that we understand little about the *benefits* provided by buildings for those that invest capital in them. Without this understanding, techniques such as cost-benefit analysis are of little use. Research that looks at the benefits is under way (e.g. Leaman and Bordass, 1999; 2004; Macmillan 2006; Ulrich *et al.,* 2004) but much more is needed before we can really start appraise investment in buildings as assets rather than as artefact costs and move beyond the rhetoric of the 1:5:200 model (Evans *et al.,* 1998 – see Ive, 2006 for a trenchant critique).

2.4. Revaluing the construction process: achieving product integrity

This section will discuss two important aspects of a revalued construction process. The first is organising the process to support the revaluation of the constructed product to create *product integrity*. The second is the distribution of the value generated within the construction process amongst its stakeholders to create *process integrity*. Both product and process integrity are project specific – a high integrity constructed product is one that is appropriate to the objectives of the client financing it as negotiated with the relevant stakeholders. Therefore, the process of achieving product integrity is one of achieving appropriate intention for the project under development. Process integrity is then a function of product integrity aimed at the consummate execution of that intent (Winch 2002). Our argument, therefore, is that revaluing construction is, essentially, a project-specific process and that the more general benefits of revaluation flow from achieving, and being seen to achieve, high integrity projects.

How, then, can we organise to achieve product integrity? While there would appear to be consensus that 'good design' is desirable, there is rather less consensus about how higher standards can be achieved. The UK Commission for Architecture and the Built Environment argues that:

> Ultimately, the responsibility for delivering high-quality projects rests with the client. It is not the procurement process itself that determines the outcome. The essential ingredients are a committed client, with the right skills and an adequate budget, focused on whole life costs, with a quality designer as part of the procurement team.
>
> (2006, p. 23)

This statement raises more questions than answers, and it is worth unpacking in order to identify what some of the issues are:

- The recommendation for 'a committed client, with the right skills' is in line with the research evidence that tough customers mean good designs (Gardiner and Rothwell, 1985) and is implemented through the recommendation that clients should appoint a 'design champion' to ensure good standards of design (OGC/CABE, 2002). The remit of this champion and their relationship with the design team is, however, unclear. For instance, the Office of Government Commerce (OGC 2007a) suggests that this person need not have experience of construction projects, but they also argue that the advice of externally appointed design professionals may also be required. In either case, it is not at all clear that such a person would understand how buildings add value for clients – even many trained design professionals are unaware of this and still treat buildings as artefacts.
- The need to choose a 'quality designer' is again difficult to gainsay, but begs the question of how such a designer is to be identified and then incentivised to give of their best on the project. As design becomes more knowledge intensive requiring understanding of not only space, form and systems in an artefactual sense, but also a deeper understanding of how these affect asset value, it is likely to become more specialised with different consultants focusing on developing competencies in specific building types. However, the advocacy of strategically partnered integrated project teams (IPTs) (Strategic Forum, 2002; OGC, 2007b) would appear to assume that IPTs have general design and construction competencies applicable to all projects, rather than competencies in specific building types. Arguably, there is a real tension between including the designer within partnered IPTs and finding the most appropriate designer for the project at hand.
- A related issue to design quality is the growing trend towards performance specifications (Leaman and Bordass, 2004). There are two dimensions to this. First, during the design stage as constructed products became more tightly engineered systems, specification is likely to become more centralised to ensure internal product integrity. As the elements of the product system become more tightly coupled, much greater attention to systems integration issues which cannot be devolved to suppliers will be required to ensure accurate sub-system interfaces. Second, greater attention to whole-life costing means that there are benefits in standardising

specifications for complex sub-systems such as lifts and escalators in order to achieve economies of scale in maintenance contracts across a property portfolio, much of which will be existing, rather than allowing decisions to be made in relation to individual projects.

2.5. Revaluing the construction process: achieving process integrity

One perspective on production processes is that they are 'value systems' consisting of multiple firms connecting final consumers to the production process in which multiple firms participate (Porter, 1985). In this perspective, firms within a value system compete with each other to retain and enhance their share of overall value created in the system – what Porter calls margin. This perspective has been criticised by writers such as Ramirez (1999) who stress the importance of co-production between suppliers and buyers and the dynamics of value constellations. Where these two authorities agree is that production essentially takes place in networks of firms, although Porter does tend to stress the firm rather than the network in most of his writing. The difference between these two positions is around whether the relationships in any particular dyad in the network are seen as zero-sum relationships in that what one gains in margin the other loses or whether through collaboration the total margin can be increased as a positive sum with gains flowing to both parties.

One way of viewing the production process is that of a flow of information and materials towards the client and then through into the building in use incentivised by a flow of investment from and through the client towards the suppliers. The client is nodal in this process because it is the client who initiates the flow of information and materials that create the asset and receives the flow of revenue generated by the exploitation of the asset. While this simple picture is, of course, complicated by lags between the revenue flows and the flows of information and materials which are covered by the use of stocks of finance (loans and equity) to lubricate the process and the provision of free-at-point-of-use facilities such as public schools, the fundamental insight is that the client is the most central node in two complementary but opposite value streams – the flows of information and materials towards the client during asset creation and away from the client during asset exploitation and the financial flows towards the client during asset exploitation as revenues and away from the client during asset creation as investments.

Value can, therefore, be defined as the client's utility for the asset being created by the construction process and this utility is a function of the asset's product and process integrity. The greater the *product integrity*, the greater the utility that the asset has for the client in subsequent exploitation; the greater the *process integrity* the greater chance of the realised asset meeting the product integrity objectives within the planned budget and schedule that formed the basis of the investment appraisal for the capital allocated to the project that initiated the value stream.

None of this, however, answers the question why, if the client captures all the utility, suppliers should participate in the construction process. The answer is that the value stream flowing from the client is shared amongst all the participants in the process in rough proportion to their contribution to the process. Suppliers mobilise resources – human and material – to create the asset in return for a share in the value stream flowing from the process (Winch, 2002). The precise quantum of this share lies somewhere between the minimum input cost of supplying the resource and the client's willingness to pay for that resource. This point is established through various mixes of market co-ordination using price signals and negotiation, and determines the utility for any particular supplier of participating in the construction process. Crucial, then, for revaluing construction is how the value stream is shared between, on the one hand, the client and its suppliers as a whole, and on the other hand amongst the suppliers themselves. In other words, the key problem is the governance of the construction process (Winch, 2001). Clearly, the greater the claim in relation to their input costs that any participant can make on the value stream associated with the creation of any particular asset, the greater their returns to participating in that creation.

This argument brings us back to the competition over margin analysed by Porter complemented by the insights of Cox *et al.* (2002), who have argued that power derived from ownership of critical assets plays an important role in this competition. However, these zero-sum perspectives on the distribution of the value stream do not take into account the role of co-production (Ramirez, 1999) in the construction process. In other words, the total value created through the process can be enhanced through collaborative working, and, therefore schemes for allocating shares in the value stream need to incentivise collaborative working if value is to be maximised. However, collaboration also has costs. Market prices are a very efficient way of co-ordinating production (Milgrom and Roberts, 1992), and co-ordination by managerial means – through a visible rather than invisible hand – has attendant costs. Crucial to collaborative working, therefore, is an assessment of whether the benefits of using managerial co-ordination for any particular transaction are likely to outweigh the costs of that co-ordination. The frameworks available for addressing this question tend to be derived from the work of Williamson (1975) and place the emphasis variously upon the importance of uncertainty, asset specificity and transaction frequency in the particular relationship under examination (Cox *et al.*, 2002; Winch, 2002) with growing attention to the 'hold-up' problem caused by process specificities and the attendant switching costs (Chang and Ive, 2007; Winch, 2006).

Advocacy of collaborative working between the parties is now widespread due to the importance of co-production for the creation of value (e.g. Bennett and Jayes, 1998), but as Cox and Ireland (2006) argue, much of it has not understood the specificities of the contribution of different suppliers to the process. The returns to collaboration will tend to be higher where there are uncertainties around the nature of the contribution, particularly where that contribution is relatively frequent and can, therefore, bear the weight of investment in collaborative governance arrangements. Holders of *critical assets* (Cox *et al.*, 2002) will have the power to take a disproportionate slice of the value stream compared to others by charging a premium for their participation in the construction process; a collaborative approach is not likely to change this underlying power dynamic and client resources may be better invested in increasing the substitutability of those critical assets and thereby reducing their owner's power rather than attempting collaboration.

The share of the value stream obtained by any particular firm will be partly passed on to firms further away from the client with the balance shared between the owners of that firm and their managers on the one hand, and those employed to make the required contribution to the construction process on the other. This is the standard economic definition of value added, but most economic analyses do not take into account its distributional aspects. Similar principles apply to the employment contract that determines the share going to labour (including nominally self-employed labour) as to the formation of business-to-business contracts within the project coalition (Winch, 2002). Again, power relations are important with greater shares of the value stream going to well-positioned work groups within the process (Frenkel and Martin, 1986). Here, too, a collaborative approach is widely advocated, while the construction industry as a whole tends to take a 'hard' rather than 'soft' approach to the motivation and development of human resources (Druker and White, 1996; Loosemore *et al.*, 2003) framed by a human capital perspective (Clarke, 2006).

Much of the current work in construction process improvement focuses on the development and implementation of new tools and techniques, but the argument here is that the key to the effective creation of value in the construction process is the alignment of incentives so that all the members of the project coalition are motivated to participate consummately rather than perfunctorily in the co-production of the built environment. Presently, much of the work in this area makes bland generalised statements on the benefits of partnering (e.g. Bennett and Jayes, 1998) or respect for people (Strategic Forum, 2002) rather than developing an understanding of how different participants in the construction process are most appropriately incentivised to give of their best in creating the built environment.

2.6. Conclusions

There is a long history of characterising the construction sector as backward – see Ball (1988) for an historical review, and Woudhuysen and Abley (2004) for a contemporary restatement. This is not the position held here. At its best, the construction sector displays a sophisticated attempt to address difficult production problems in a way that draws respect from other supposedly high-tech sectors such as IT (e.g. Standish Group, 1995), and has been cited by influential commentators as a model of the contemporary organisation of production (cf. Winch 1994). Just as the automobile sector – defined in the same was as construction (Winch, 2003) – has its incompetent back-street garages and dodgy car salesmen, the construction sector has its weaknesses and certainly has room for improvement. Revaluing construction is about presenting this achievement positively while honestly addressing the weaknesses of the sector. This short chapter has focused on some of the issues for the construction process around two areas where further research and practice development is needed – organising to achieve product integrity and developing appropriate incentive alignment for process integrity. Left untouched are large areas of the revaluing agenda such as the stakeholder dimension where value creation for one stakeholder such as the client can involve value destruction for other such as local residents, and the sustainability dimension where the ethical and economic definitions of value come together. However, it is hoped that some progress has been made in developing a framework for understanding better constructed products as assets rather than artefacts as a function of process and product integrity.

Acknowledgements

I am very grateful to Michael Ball and Graham Ive for their comments on this chapter. In particular, Graham Ive provided detailed feedback on the economic concepts deployed. Neither, of course, bears any responsibility for the final text.

References

Ball, M. (1988) *Rebuilding Construction*. London, Routledge.

Barber, W.J. (1967) *A History of Economic Thought*. Harmondsworth, Penguin.

Barrett, P. (2007) Revaluing construction: a holistic model. *Building Research and Information* **35**: 268–286.

Bennett, J. and Jayes, S. (1998) *The Seven Pillars of Partnering: A Guide to Second Generation Partnering*. London, Thomas Telford.

Boussabaine, A. and Kirkham, R. (2004) *Whole Life-Cycle Costing: Risk and Responses*. Oxford, UK, Blackwell.

Clark, K. and Fujimoto, T. (1991) *Product Development Performance: Strategy Organization and Management in the World Auto Industry*. Boston, Harvard Business School Press.

Clarke, L. (2006) Valuing labour. *Building Research and Information* **34**: 246–256.

Chang, CY and Ive, G. (2007) Reversal of bargaining power in construction projects: meaning, existence and implications. *Construction Management and Economics* **25**: 845–856.

Commission for Architecture and the Built Environment (2006) *Better Public Building*. London, CABE.

Cox, A. and Ireland, C. (2006) Strategic purchasing and supply chain management in the project environment – theory and practice. In: *Commercial Management of Projects: Defining the Discipline* (eds D. Lowe and R. Leiringer). Oxford, UK, Blackwell.

Cox, A., Ireland, P., Lonsdale, C., Sanderson, J. and Watson, G. (2002) *Supply Chains Markets and Power: Mapping Buyer and Supplier Power Regimes*. London, Routledge.

Davenport, T.H. (1993) *Process Innovation: Reengineering Work Through Information Technology*. Boston, Harvard Business School Press.

Dickson, M. (2004) Achieving quality in building design by intention. In: *Designing Better Buildings* (ed S. Macmillan). London, Spon.

Druker, J. and White, G. (1996) *Managing People in Construction*. London, Institute of Personnel and Development.

Egan, J. (1998) *Rethinking Construction*. London, Department of the Environment, Transport and the Regions.

Evans, R., Haryott, R., Haste, N. and Jones, A. (1998) *The Long Term Costs of Owning and Using Buildings*. London, Royal Academy of Engineering.

Frenkel, S. and Martin, G. (1986) Managing labour on a large construction site. *Industrial Relations Journal* **17**: 141–157.

Gann, D., Salter, A.J. and Whyte, J.K. (2003) The design quality indicator as a tool for thinking. *Building Research and Information* **31**: 318–333.

Gardiner, P. and Rothwell, R. (1985) Tough customers: good designs. *Design Studies* **6**: 7–17.

Hammer, M. (1990) Reengineering work: don't automate, obliterate. *Harvard Business Review* July–August: 104–112.

Ive, G. (2006) Re-examining the cost and value ratios of owning and occupying buildings. *Building Research and Information* **34**: 230–245.

Leaman, A. and Bordass, B. (1999) Productivity in buildings: the 'killer' variables. *Building Research and Information* **27**: 4–19.

Leaman, A. and Bordass, B. (2004) Flexibility and adaptability. In: *Designing Better Buildings* (ed S. Macmillan). London, Spon.

Loosemore, M., Dainty, A. and Lingard, H. (2003) *Human Resource Management in Construction Projects; Strategic and Operational Approaches*. London, Spon.

Macmillan, S. (2006) Added value of good design. *Building Research and Information* **34**: 257–271.

Milgrom, P. and Roberts, J. (1992) *Economics, Organization and Management.* Upper Saddle River, NJ, Prentice Hall.

Office of Government Commerce (2007a) *Design Quality.* London, OGC.

Office of Government Commerce (2007b) *The Integrated Project Team: Teamworking and Partnering.* London, OGC.

Office of Government Commerce/Commission for Architecture and the Built Environment (2002) *Improving Standards of Design in the Procurement of Public Buildings.* London, OGC/CABE.

O'Hare, M. (1997) Attention, value and exchange. In: *Center 10/ Value* (eds M. Benedikt, S. Najaran and J. Stone). Austin, School of Architecture, The University of Texas.

Porter, M.E. (1985) *Competitive Advantage: Creating and Sustaining Superior Performance.* New York, Free Press.

Prasad, S. (2004) Inclusive maps. In: *Designing Better Buildings* (ed S. Macmillan). London, Spon.

Ramirez, R. (1999) Value co-production: intellectual origins and implications for practice and research. *Strategic Management Journal* **20**: 49–65.

Rouse, J. (2004) Measuring value or only cost: the need for new valuation methods. In: *Designing better buildings* (ed S. Macmillan). London, Spon.

Smith, A. (1970) *The Wealth of Nations.* Harmondsworth, Penguin.

Spencer, N. and Winch, G.M. (2002) *How Buildings Add Value for Clients.* London, Thomas Telford.

Standish Group (1995) *The Standish Group Report: Chaos.* Boston, The Standish Group.

Strategic Forum (2002) *Accelerating Change.* London, Rethinking Construction.

Thompson, D.S., Austin, S.A., Devine-Wright, H. and Mills, G. (2003) Managing quality and value in design. *Building Research and Information.* **31**: 334–345.

Ulrich, R., Quan, X., Zimring, C., Joseph, A. and Choudhary, R. (2004) *The Role of the Physical Environment in the Hospital of the 21st Century: A Once-in-a-Lifetime Opportunity.* Concord, CA, The Center for Health Design.

Williamson, O.E. (1975) *Markets and Hierarchies: Analysis and Anti-Trust Implications.* New York, Free Press.

Winch, G.M. (1994) The search for flexibility: the case of the construction industry *work. Employment and Society* **8**: 593–606.

Winch, G.M. (2001) Governing the project process: a conceptual framework. *Construction Management and Economics.* **19**: 799–808.

Winch, G.M. (2002) *Managing Construction Projects: An Information Processing Approach.* Oxford, UK, Blackwell Science.

Winch, G.M. (2003) How innovative is construction? Comparing aggregated data on construction innovation and other sectors – a case of apples and pears. *Construction Management and Economics.* **21**: 651–654.

Winch, G.M. (2006) The governance of project coalitions: towards a research agenda. In: *Commercial Management of Complex Projects: Defining the Discipline* (eds D. Lowe and R. Leiringer). Oxford, UK, Blackwell, pp. 324–343.

Womack, J.P. and Jones, D. (1996) *Lean Thinking; Banish Waste and Create Wealth in Your Corporation.* New York, Simon and Schuster.

Womack, J.P. Jones, D.T. and Roos, D. (1990) *The Machine that Changed the World.* New York, Rawson Associates.

Woudhuysen, J. and Abley, I. (2004) *Why Is Construction So Backward?.* Chichester, Wiley-Academy.

Wright, R.N., Rosenfeld, A.H. and Fowell, A.J. (1995) *National Planning for Construction and Building R & D.* Gaithersburg, MD, National Institute of Standards and Technology.

3 Is the client really part of the team? A contemporary policy perspective on Latham/Egan

John Hobson and Kenneth Treadaway

3.1. Introduction

It has been a mantra since Sir Michael Latham published his report in the UK 'Constructing the Team' (Latham, 1994) that the client is part of the construction team. This is not a new concept. Government sponsored reports over the years, such as the Banwell Report of 1964 (Banwell, 1964), have examined the issue of improving construction industry performance through increasing focus on the client, with surprisingly similar but equally ineffective conclusions. But the John Major government's concerns about government/industry relations led to the serious attack on the issues represented by the Latham review and continued in the Labour era by Sir John Egan's Task Force (Egan, 1998). The Latham process not only produced insights into the client role, but also led to the creation of centralised machinery (the Construction Industry Board), which involved an explicit role for client representatives. The Egan Task Force with its dominant client membership, built on this process with a focus on demonstration projects, a ginger group the Movement for Innovation (M4I), and performance measurement using key performance indicators throughout the construction process, mirroring Egan's experience in the motor industry.

The authors do not question the validity, at the time, of the client-focussed insight. For some years this was a liberating and stimulating concept, helping to eliminate some of the worst intra-industry confrontations and enabling the concepts of teamwork, partnering and collaboration to take root and grow. But what does it actually mean, and is it still (a) true and (b) helpful?

3.2. Is the client knowledgeable?

There is no reason why a construction client should be expected to know anything about construction. This is a professional industry. As clients of doctors or dentists we do not expect to have to help with health care, except as patients giving information about our illness; as airline passengers we do not expect to need a grounding in aircraft design or aviation theory. So why are we expected to help the construction industry with its internal workings? The client operates from outside the industry and buys in its services. He/she is entitled to expect a professional service.

Some clients have made themselves not only members but almost leaders of the team. The British Airports Authority (BAA) PLC is an obvious example. There are special reasons for this. BAA acts as an intermediary for many of its clients, the airlines for example, and this puts BAA in the special position of having to understand both its suppliers and its customers. Some of the results have been good (e.g. Terminal 5 (T5) at London Heathrow Airport), but it is said that the bureaucracy associated with preferred supplier status is extensive and can result in a costly duplication of effort. Could this stem from a basic lack of trust in the supply side? Is this teamwork? Other commercial clients have effectively eliminated formal contracts, but they are clients requiring almost identical products widely built, the prime example being McDonalds Restaurants. Many, however, remain one-off construction procurers or even avoid the complexities of procurement through leasing. They remain clients of the construction industry, but have quite different perspectives on construction than those clients closely associated with construction delivery.

3.3. Client responsibilities

Some client responsibilities are clear. Clients must be experts on their own context/ business. And they must be able to enunciate, on paper or otherwise, the policy outcomes and outputs that they are looking for from the project. One of the best features of the government's Private Finance Initiative (PFI) is the importance it attaches to a precise client generated output based brief; this is a very necessary discipline. Furthermore, it is only common prudence for clients to keep a track of progress both in timeliness and budgetary performance. Other involvements are not so clear. For example, clients should not have to concern themselves with the inner workings of the supply side team. Indeed as amateurs they are likely to cause much more heat than light. (Perhaps doctors feel much the same about hypochondriacs who come armed with medical information taken off the internet!).

It is very convenient for the supply side of the industry to seek to shuffle off responsibility for performance improvement to the client. They allege poor performance review, changes of specification and lack of cost control. However, many of these shortcomings could be mitigated by a genuinely professional supply side that had the client's interests rather than its own profits at heart. A special case of this is government, who are a major construction client as well as industry sponsor and regulator. We now turn to these complex but inter linked roles.

3.4. The role of government

Government comes in many fragmented manifestations and is driven by both political and operational objectives, so government as client is always likely to speak with at least two voices and very likely more. (The Football Association's Wembley Stadium redevelopment and the development for the London (UK) Olympics in 2012 are instructive here, as compared with the building of Arsenal Football Club's new Emirates stadium.) The UK government had success with the Ministry of Defence's (MOD) 'Building Down Barriers' project (Holti *et al.*, 1999), but had serious problems with the Millennium Dome (where the client failed to define the ultimate purpose of the technically outstanding development) and other major projects like the Jubilee Line Extension (where mechanical and electrical services and train control systems added significantly to the complexity of the project as requirements changed). Attempts at repeated projects in the health, defence and education field (e.g. National Health Service (NHS) Local Improvement Finance Trust (LIFT), 2007; NHS Procure 21, 2007; Building schools for the future, 2007) have proved challenging and remain ongoing issues given the government's wish for continued heavy investment in these areas. (see Chapter 21 for a case study of a major capital development programme, the Manchester, Salford and Trafford Local Improvement Finance Trust.)

Because the UK government knows itself to be fragmented, it seeks to provide a co-ordinated face to the industry. In the past the Ministry of Public Buildings and Works – later the Property Services Agency (PSA) – pulled together construction and design expertise relating to all government departments and provided a service to them. This system changed to a department sponsored service on the grounds that the PSA could not be properly accountable for other departments' spending. This is a powerful argument, and resulted in the Office of Government Commerce (OGC), an office of the UK Treasury, taking over procurement coordination. It now sends out the circulars and centralised procurement advice, which used to be the preserve of the highly professional dominated PSA. But in the changes much in-house construction expertise has been lost. It is said that there is only one person with experience of construction in the OGC. That cannot be right. And the balance of power remains with the big spending departments: MOD, Department for Education and Skills (DfES), Department of Health, Highways Agency (HA), etc.

The UK government can also tilt the playing field through its taxation, fiscal and regulatory powers. Experience suggests that it will always be a poor team player, with its own political and economic agendas. Observe the difficulty with discerning the economic objectives behind the aggregates tax or the planning gain supplement; yet fiscal instruments are probably the most effective counter to global warming. The Treasury is expert at influencing an industry through uncertainty as to the angle of attack – an approach weakened by the separation of the OGC.

The situation is exacerbated when, as at present, the industry sponsorship profile is low. Sponsorship is an elusive concept, but the 10 years of the Latham/Egan period demonstrated the effectiveness in achieving change of a powerful sponsor department (Department of the Environment/Department of Environment, Transport and the Regions (DOE/DETR)) with senior, knowledgeable and respected ministers. In present circumstances where these elements are missing it is unlikely that government will instigate major change programmes. History demonstrates that little else will move this industry sector.

In summary, government is not the sure-fire change agent that it is sometimes seen as present circumstances militate against major government initiatives. The industry's own complaints in the past about 'initiative-itis' have perhaps come back to haunt it.

3.5. Small clients

Domestic clients genuinely do not have the time or knowledge to act as team members. There is no reason why they should, but they do need to know what they want. A client's partner can cause mayhem with a contract by making changes during the day while the other half is at work! Few domestic clients draw up written contracts even though standard forms are available. A truly professional industry would provide this kind of inexpert client (who probably provides something like 50% of the industry's workload by value including maintenance) with a reliable, trustworthy service, giving high design standards, certainty of performance and certainty of value for money. We expect our cars to work without our supervising production and maintenance. We expect butchers to know how to cut up meat and charge us a fair price. Why not the same for building production and maintenance? But while many small builders continue to provide a professional and expert service with a highly integrated supply chain much experience since the Pharaohs shows a high degree of customer dissatisfaction. No doubt the Pyramids had to be resited several times before their astronomical properties could be made to work!

3.6. Client representation at policy level

For too long the supply side of this industry has ducked its own responsibility by delegating responsibility for its performance to either government or 'the client'. To its undying credit, for 10 years the client entity has shouldered this inappropriate burden and, with the enormous assistance of individual clients such as Sir John Egan and Sir Stuart Lipton (Stanhope PLC), has achieved substantial change. It could not have done this without the active participation of government as sponsor and client; and, without strong Treasury complicity, which forced the industry to take notice for fear of what might happen if they did not.

But the client population has proved impossible to organise as a collective, despite the best efforts of many. It is too big and too varied. It encompasses the vast majority of the population at plumbing and small work level (including DIY), ranging through to oil majors and multi-nationals. No one body can meaningfully represent such a range of interests.

It is now time for the supply side to take on the responsibility itself for delivering professional standards of work to meet its clients' objectives and aspirations through provision of the service they have every right to expect. To do this, however, the suppliers must gain a better understanding of their 'clients', and there are many and varied manifestations of 'clients' each of whom has different motivations, aspirations and commercial/social/emotional drivers. Part of this must be down to trade associations and professional bodies, all of whom claim to have the best interests of the industry at heart but are in fact primarily (and rightly) motivated by the interests of their members.

History is again instructive. Efforts were made as part of Latham implementation, and the setting up of the Construction Industry Board (CIB), to produce a representative client body, the Construction Clients Forum (CCF), based on the highly successful (in lobbying terms) British Property Federation (BPF) and the Construction Round Table (CRT). The federation that emerged represented most elements of clients' interests, including public and private sector clients. It did, however, have one failing in that there were few individual client members, CRT and BPF acting as surrogates for many of the major private sector clients. Because of this CCF was always in danger of producing 'one size fits all' advice, which had to be tailored to meet individual client's interests. It also had a major problem of representing the small and occasional clients who were essentially fragmented in their interests. The consequence was tension between the different client interests, an endemic problem when clients are bulked together as a whole. Whilst attempts were made to rectify these faults in its successor body (the Confederation of Construction Clients, (CCC)) these same tensions and difficulties remained.

In the end, client uncertainties led to the disbandment of the CIB without any adequate central industry body replacement at strategic level; a real step backwards. However, Constructing Excellence has now been established combining M4I and the Best Practice Programme, united with Be[1], and working with the Strategic Forum for Construction (www.strategicforum.org.uk) to fill the gap left by the CIB and, interestingly, the current grouping, Constructing Excellence in the Built Environment (CE) includes client representation through its Construction Clients Group with its membership drawn both from industry bodies and individual clients (not unlike CCF). These events are comprehensively chronicled from a client perspective in Adamson and Pollington (2006).

But there have been successes. The BPF has already been mentioned, as has the CRT. CRT was a small group of expert clients with parallel interests founded in 1992 under the chairmanship of Sir Christopher Foster. CRT provided a broad representation of client interests in the absence of any formal representation in construction (rather than property management and ownership).

When CRT was established its members agreed that there would be only one representative from any particular segment of industry. This ensured both wide representation and the opportunity to discuss, in confidence, sensitive business-specific interests, which members might not wish to raise in a more public arena. While this model worked well in the initial stages members felt that they were not able to make the impact they wished in the wider construction arena, in particular, in influencing industry change to address their needs and business aspirations, and in policy discussion with government. Accordingly the organisation was revamped in 1997 with the publication of its *Agenda for Change* (CRT, 1998).

Agenda for Change was a statement of the very high standards CRT members sought in the planning, delivery, operation and performance of their facilities. It focused on designing members' facilities, the trading environment in which they operated with the supply side of the industry and the construction delivery process, which members believed were critical to improving their performance. *Agenda for Change* was supported by a programme of technical development and demonstration projects that were implemented in collaboration with the industry and which were tested by measuring progress in improving performance.

Bearing in mind the CRT membership represented some 25% of the purchasing power of the construction industry; CRT's proposals were welcomed by the industry at large. *Agenda for Change* also had significant effect on government, which recognised the principles as being of wider importance in changing the way in which the construction industry operated. In parallel, the government had set up the Egan Task Force to develop a framework in which all elements of the industry, clients, suppliers, designers and consultants might work together for the betterment of the industry at large. The outcome, *Rethinking Construction,* contained many of the principles outlined in *Agenda for Change,* (not surprisingly given that the same group of people was involved in both processes) and in particular took forward the principle of improving through demonstration. CRT members enthusiastically took up the twin challenges of demonstration and measurement. In so doing, the CRT had achieved its primary goal of precipitating the principles of change and improvement in the industry to meet clients' needs and wishes. Its main driving force was concentrated in support of the M4I (the main Egan change mechanism) and while it continued for some little time after the establishment of M4I it lost its core impetus. The fragmentation of interests of even these major clients surfaced through the differences between clients who constructed and owned, owned or simply operated as business tools their constructed assets. What was left, however, was an enduring interest in the concept of construction as a business asset creator rather than merely a delivery process ancillary to the core business of the client.

3.7. Conclusion

The above discussion leads to the conclusion that the 'client' as one body is wholly inappropriate and attempts to organise clients in a unified way are doomed to failure unless recognition of the diverse interests of individual clients can be embraced within the single 'client' body. Individual client's interests within the wider body need to be recognised and catered for and as a start a classification of interests and types of clients is necessary. With this the understanding of the client, client needs and targeted action can be better undertaken. It may be the absence of this essentially academic insight that lies at the heart of our repeated failure to establish a lasting and credible client voice.

The University of Salford, UK, is currently engaged in formulating a taxonomy of clients. This is a welcome first step, which might lead to a better understanding in this essential area, stratification of the client population, and translation into meaningful groupings of clients, which could then be formed into a tighter or looser – according to taste – client federation. Such a federation would be well placed to redefine the place of the 21st century client in the team. And it would do so from a position of genuine insight, influence and political power.

Note

[1] Be: 'Collaborating for the Built Environment', a merger of the Design Build Foundation (DBF) and The Reading Construction Forum (RCF).

References

Adamson, D. and Pollington, A. (2006) *Change in the Construction Industry.* London, Routledge.

Banwell, Sir H. (1964) *The Placing and Management of Contracts for Building and Civil Engineering Work.* London, Ministry of Public Building and Works, HMSO.

Building Schools for the Future (2007). Available at: http://www.bsf.gov.uk/ [Accessed December 2007]

Construction Round Table (CRT) (1998) *Agenda for Change.* Garston, Herts, BRE.

Egan, J. (1998) *Rethinking Construction: Report of the Construction Task Force to the Deputy Prime Minister, John Prescott, on the Scope for Improving the Quality and Efficiency of UK Construction.* London, DETR, HMSO.

Holti, R., Nicolini, D. and Smalley, M. (1999). *Building Down Barriers, Prime Contractor Handbook for Supply Chain Management*. London, Ministry of Defence.

Latham, M. (1994) *Constructing the Team: Final Report on Joint Review of Procurement and Contractual Arrangements in the UK Construction Industry*. London, HMSO.

NHS Local Improvement Finance Trust (LIFT) (2007). Available at http://www.dh.gov.uk/en/Publicationsandstatistics/Publications/PublicationsPolicyAndGuidance/DH_4010358 [Accessed December 2007].

NHS Procure21 (2007). Available at http://www.nhs-procure21.gov.uk [Accessed December 2007].

4   Enabling clients to be professional

Roger Courtney

4.1. Introduction

Drawing on inputs to the International Construction Clients Forum (ICCF) and to Revaluing Construction conferences, this chapter presents a concept of the 'professional client', defining the client role in terms of the relationship between the client and the stakeholders in the project. It discusses how the current consensus on best client practice has required clients to innovate and to stimulate innovation in the project partners and considers the consequent knowledge gaps and research issues. Finally, it notes that this consensus is drawn from experience of projects in construction business systems that broadly follow an Anglo-Saxon model and suggests that the client function in other construction business systems should be studied.

4.2. Defining the client role

The ICCF[1] is a grouping of client interests that has come together under the auspices of International Council for Research and Innovation in Building and Construction (CIB). The Forum acts as a network through which members may seek information and advice on client-related issues. Its meetings have been attended by national client associations (such as those established in Sweden and Denmark), bodies founded to promote better practices in construction (e.g. the Construction Industry Development Board in South Africa) and individual client organisations, both in the public and private sectors. The inaugural meeting of the Forum was held in The Hague in September 2004 and a second meeting took place in Port Elizabeth in October 2005.

The Forum defines the client role through its relationship with the supply chain and the various stakeholders who have interests in the final constructed output. The principal relationships are summarised below:

4.2.1. Owners/financiers

Clients represent owners' interests through being concerned to achieve a final output that will support the owner's business or other corporate objectives, and will add value to the owner organisation. At the pre-project phase, the owner will need to consider whether their business objectives are best met through construction, or could be achieved through another form of management initiative such as an organisational change; the client organisation will be well placed to contribute to that decision. The client will go on to participate in the project process and will take delivery of the completed facility, to oversee its commissioning and its acceptance by users and, ideally, to arrange for evaluations of performance in order to inform future projects.

4.2.2. Users

Clients are responsible for identifying and communicating users' requirements to supply interests, through the preparation of a 'brief' or equivalent statement of needs. Ideally, this statement should anticipate future needs since to be sustainable the building or other facility will need to function effectively for future generations of users.

Clients may be able to interact directly with current users, or may need to draw on experience to project the needs of unknown users. Some tools for facilitating user and other stakeholder input to the design process have been developed, but these are in their infancy. IT-based communication concepts such as virtual reality offer the promise of much more effective communication routes but again require development before they can be commonly employed.

4.2.3. Society

The built environment is a common good – everyone is influenced by it. Hence 'society', acting through public bodies and interest groups, is a key stakeholder in any project. Society's influence is exercised through regulatory procedures such as planning controls and it is the role of the client to be sensitive to these requirements and to work with relevant bodies, and with designers, to achieve an optimum solution for all stakeholders.

4.2.4. Suppliers

Clients have great influence on relationships and project delivery structures in construction. They exert this through their selection of procurement systems and associated financing and contractual arrangements, through the criteria that they employ in selecting designers and contractors and through the quality of their interactions with the supply chain.

In summary, therefore, the client is, throughout the project, the interface between the supply chain and other stakeholders. This formulation usefully distinguishes the client role from those of the owner, user or financier, although in many cases the client organisation will sit within a larger body with units that fulfil some or all of these roles. The client organisation is then required to relate to these other functions as stakeholders in the project, and to seek to satisfy their requirements.

4.3. Client leadership and the 'professional client'

In the past 10–15 years, the performance of construction sector has come under scrutiny in many countries and the creation of the ICCF is one manifestation of an international search for improved value from expenditure on construction projects. The drive to improve performance has stemmed from different causes: the desire to reduce costs, inappropriate behaviour by construction interests, poor labour relations, etc. Some countries have undertaken formal reviews of the industry and have initiated national reform or improvement programmes.[2]

A common conclusion of such reviews has been that successful projects are not just the result of good performance by the supply side; the policies, practices and behaviour of clients for construction projects have a very large influence on their ultimate outcome. The way that clients set out their requirements and procure their projects determines relationships and practices within the supply sector. The value that clients obtain from expenditures on construction is, therefore, directly related to their own ability to operate successfully as clients.

From the many publications, presentations and project reports that have been produced since around 1990, a global consensus has emerged on client practices that lead to superior outcomes in at least the larger and more complex projects, which in the past have often been delivered late and over budget. This consensus focuses on the adoption of policies and practices that will engender and reward collaboration among all parties and will generate mutual commitment to the success of the project. Key to the creation of such a collaborative environment, and a radical break from traditional practice, is the early appointment of a project team that includes all the principal parties. This is explicitly incorporated, for example, in guidance from the Office of Government Commerce (OGC, 2003), which sets procurement policy for UK government bodies: 'An integrated team should be appointed to carry out the project' (p. 5).

Integration may be a product of the contract strategy adopted (e.g. the use of design-build or another form of contract which has the effect of creating a single-point responsibility for project delivery) or may be promoted by the behaviour of the client following the appointment of the principal partners to the project. The use of integrated forms of contract is strongly advocated by OGC, 2003, p. 5): 'Traditional contract strategies, where the design and construction are provided separately, should only be used where it can be clearly demonstrated that this approach will provide better value for money than the preferred integrated procurement routes'. However, elsewhere integration is encouraged within the context of conventional contract forms. For example, contractors in Denmark are appointed early in 'partnering' contracts on terms that cover their input to the project design (Agency for Enterprise and Housing, 2004) then, once the design and the price are settled, a contract for construction is agreed.

Positive client leadership, which sets and maintains the prevailing culture of the project, is therefore a pre-requisite of success. Client leadership is not to be confused with client dominance; this would breach the principle of collaboration. Client leadership is a fine balance between, on the one hand, exerting influence and taking decisions – recognising that the client has ultimate responsibility for the project – and, on the other, being open to ideas from all sources, including those that may question some key aspects of the project, in order to achieve the best outcome. Client leadership may be characterised as follows:

- Clarity over the client's expectations both for their own staff and operations and for those of their partners in the project.
- Active participation of the client throughout the project. The client is a member (and may be the formal leader) of the project team.

- Consistency and rigour in the application of monitoring systems to ensure that expectations are met and appropriate mechanisms for handling any shortcomings in performance.
- Open communications before and during the project, to agree objectives and targets.

Recognising the importance of client leadership in construction projects, and the way in which this must draw upon experience, knowledge and expertise, the ICCF has developed the concept of the 'professional client' and formulated a 'Vision' for construction centred on this concept. The Vision is that: 'Construction projects are routinely successful because they are inspired and led by clients who act in a professional manner'.

The Forum is now exploring how the professionalism of clients may be enhanced. One way is through education. The Swedish Construction Clients Forum (SCCF) (Byggherre forum) (SCCF, 2006) a prominent member of the ICCF, has made the Vision operational by stimulating the development of a graduate (MSc) course focussed on the knowledge and expertise requirement by clients, to be delivered by the four Swedish technical Universities starting in January 2007.

4.4. 'Value' and the client role

A significant recent development in the way that clients look at their role has been the introduction of 'value' concepts.[3] These influence both the eventual output and the method of procurement. They have been stimulated, in part, by the rise of new funding and contractual structures, notably Public–Private Partnerships, which have focussed attention on the operational performance and life-time costs of the completed facility in place of the previous concentration on initial performance and costs.

4.4.1. Value and buildings

It has, of course, been accepted for centuries that buildings are appreciated (or, in a general sense, have value) for reasons that are not connected to their ability to meet the needs of their direct users. The classical Vitruvian formulation of the aims of architectural design – 'commodity, firmness and delight' – is an expression of the desired properties of utility, durability and aesthetics in the final building.

One analysis (Spencer and Winch, 2002) of the 'value' of buildings has expressed the components of value in terms of the following:

- Spatial quality – the way that the arrangements of space within the building support the functions within it and its impact on the surrounding urban area.
- Indoor environmental quality – the building's impact on its users and its influence on their effectiveness.
- Symbolism – the way the building conveys the values of its owners and users.
- Financial value – how these other aspects are effected in monetary terms.

The first of these – the impact of buildings on the activities taking place within and around them – is a subject of growing interest, reflecting recognition that the economic cost of those activities over the lifetime of the building is likely to exceed its initial and operation costs by a large margin. And in the urban context, the impact of building on its surrounding area, because of both its functional and aesthetic characteristics may result in considerable enhancement in business activity and property values – the effect of Guggenheim Museum on the port area of Bilbao, Spain is a classic example. Hence value-based procurement, in relation to the actual building or works, involves assessment of the value provided to a large range of stakeholders, through both formal evaluations and more general consultation processes.

At the present state of knowledge, it may not be possible to express all these aspects of value in a common currency, but some current practices are steps towards that. In particular, the use of whole-life costs rather than merely initial costs in the assessment of the cost of the works ensures that operational (and, potentially, disposal) issues are taken into account and may be set against operational benefits. The OGC (2003) guidance referred to earlier is explicit on this point: 'procurement decisions on construction projects should always be taken on the basis of value for money over the life of the facility and not the initial capital cost' (p. 2). Formal systems for assessing building performance and whole-life costs are now developed sufficiently to be incorporated in national and international standards.[4]

Further, the practice of stating client requirements in performance or outcome terms will promote the achievement of higher value. By not restricting suppliers to a pre-defined solution, different means of arriving at the desired goals will be encouraged, particularly when the commercial relationship between client and suppliers includes incentives for higher performance and reduced costs. CIB Working Commission W60 has been an international focus for the development of performance-based procurement. As understanding of the impact of buildings and facilities on the activities within them increases, the ability to express requirements in terms of outcomes (e.g. for health facilities, impacts on patients) will correspondingly increase.

4.4.2. Value and suppliers

The use of 'value' concepts in the section of suppliers (and particularly contractors) again involves the use of a much broader range of criteria than merely the traditional 'lowest cost' approach. Typically, cost will be a significant factor in the decision, but a range of others will be taken into account, the final decision being informed by a weighted combination of evaluations against the different criteria: health and safety record, environmental performance, past delivery record, investment in training, etc. Some of these additional factors relate to satisfying the 'value' goals of the wider community. For instance, in South Africa socio-political factors have influenced procurement policies, with the aim of addressing the legacy of social inequality in that country.[5]

An alternative to the use of criteria tailored to individual projects is the use of a pre-qualification scheme. This has the effect of introducing non-price factors into the selection process, by requiring prospective suppliers to satisfy a range of criteria. The final choice, from a tender list of pre-qualified suppliers, may then be more influenced by price. Such schemes have been used for many years, particularly to ensure that firms have appropriate technical and financial capability for the works involved. However, with increasing emphasis on the ability of project participants to work together in harmony, some schemes are now seeking to assess the firm's attitudes to collaboration and their ability to engage with others in a team.[6]

4.5. Clients and innovative behaviours – some issues

The non-traditional procurement practices outlined above, and the introduction of value concepts in clients' appraisal and procurement of projects, are two ways in which leading clients have demonstrated innovative behaviour. Clients can insist on this, but to have real impact these changes require corresponding changes throughout the project team. By changing their modes of operation and the way they express their requirements, leading clients therefore have the power to cause consequential change in their attitudes and practices in their supply partners and in the final delivered output – this is undoubtedly a case of *clients driving innovation*.

However, while the general directions of change required in the client role and behaviour are clear, many issues remain to the investigated, and because these relate to interactions among individuals and organisations, there may be no universal solutions. Local culture and practice may well influence the optimum solution in each country. These issues provide themes for current and future research. The discussion below outlines some of them.

4.5.1. Stimulating and maintaining change

The changes required if clients are to demonstrate leadership, while at the same time promoting collaboration, must extend throughout the client organisation. Traditional attitudes and behaviours have to be addressed, and new approaches may seem to expose staff to more risks and to impose new responsibilities on them. Top-level commitment to change is essential, but not sufficient. Training and 'change' programmes are required, with recognition and reward systems amended to be consistent with new corporate aims. As an illustration of the challenge, recent reviews of client–supplier relationships in the UK (Business Vantage, 2004a, 2004b) have shown that despite the effort put in to change construction cultures in the UK over the past decade, both sides are dissatisfied with progress; delivery does not match the top-level policies. Hence:

- Which change strategies are most effective in developing client leadership behaviours and capabilities? How do these relate to the starting culture?
- How can the impetus for change be maintained?
- What are the most relevant metrics for assessing progress?
- Is there a difference between the public and private sectors, in the ability to adopt client leadership?

4.5.2. Defining responsibilities

When clients adopt a new, leadership role, there are implications not only for them but for other project participants, notably for architects who have traditionally acted on behalf of the client. The Danish Association of Construction Clients[7] has, for example, advised that there is a risk that clients – perhaps unwittingly – will take on responsibilities that have previously fallen to the supply side:

- What guidelines can be developed to assist clients in finding the balance between exerting a proper leadership role and taking on too great a share of the project risks?
- How should the relationships between the client and their professional advisers be structured to be consistent with client leadership?

4.5.3. Creating the value proposition

The use of 'value' as the basis for procurement decisions implies that the overall value provided by different options can be compared. Some elements of the value equation (such as initial cost) are capable of estimation with reasonable accuracy; others (e.g. cost of future maintenance) are open to more uncertainty and some (such as the way alternative designs may impact on the activities being carried out, or on the attractiveness of the local neighbourhood) are currently assessable only in very imprecise terms. Hence, there is lack of understanding of some main elements in the overall value proposition, and inadequate methodology for bringing different dimensions of value together in a common metric:

- What methodologies might be imported from other industry sectors to assist the application of wider concepts of value at the design stage of a project?
- How can evidence of different aspects of value (from post-occupancy studies, urban regeneration, capital appreciation, etc.) be brought together in a common framework?

4.5.4. Selecting the supply partners

When considering value as a basis for selecting the project team, or individual members of the team, the key task is again the combination and balancing of the price and non-price factors that will influence the decision. As before, though, some important factors (e.g. the capacity for collaborative behaviour) may be difficult to assess:

- Are there general combinations of price and non-price factors that have been found to work over a range of project types?
- What does experience teach about the effectiveness of selection processes? For example, can pre-qualification adequately cover the non-price factors or is it always necessary to include project-specific factors in the selection process?
- How can the 'soft' issues such as management style and attitudes be judged in the absence of direct experience of working with a firm?

4.5.5. Satisfying public accountability requirements

Value-based procurement, whether applied to the product or the suppliers, may result in a project with a higher initial cost than traditional processes would produce. Many of the largest construction clients are accountable to public audit bodies and operate against a long tradition of highly competitive 'lowest price' procurement. It is understandable that processes that may appear to lead to higher initial costs, or to encourage collaboration between clients and suppliers, may be viewed with some scepticism. Within the EU, for example, the latest Procurement Directives (European Commission, 2004a, 2004b) permit selection on the basis of the 'most economically advantageous' option, but a recent study for the European Commission, led by the author and colleagues at Manchester Business School (Courtney *et al.*, 2006), revealed widespread concern that lowest price was still the most widely used criterion. Hence, a strategy for demonstrating the advantages of value-based procurement, for example, through case studies of the impact of new decision criteria, is required: which approaches are most effective in promoting the use of value-based procurement in the public sector?

4.6. A global view – or a particular perspective?

The previous discussion of the client role and associated issues is founded on studies and presentations from around the world, many being the subject of presentations to the Revaluing Construction conferences sponsored by CIB in 2003 and 2005. A similar perspective on the client function and the factors that promote successful projects emerges from presentations and discussion at meetings of the Civil Engineering Research Foundation in the US. It is, however, notable that these studies originate from those countries whose construction business systems (Winch, 2000) follow an 'Anglo-Saxon' or Scandinavian model. These systems are characterised by a fragmentation of responsibilities within the supply chain and a clear separation between client and supply functions. There appears to be no comparable development of client awareness, or formulation of the client role, in countries with more integrated construction business systems such as France, Spain or Japan.

This may be a reflection of more general cultural and institutional factors, such as the close relationship between State and private functions (and the associated interchange of senior executives) in France or the collaborative culture that is characteristic of Japan. One may also speculate that perhaps it reflects an intrinsic ability of the construction business systems in those countries to deliver higher value to clients, thus reducing the stimulus for client action and the development of a professional client function. If so, there might be lessons for other construction business systems. Comparisons of the performance of national construction sectors are problematic, but with major clients now operating in global markets, there is a strong case for establishing reliable comparators. CIB Task Group 61 (Benchmarking Construction Performance Data) is addressing some aspects of this issue:

- How is the client function interpreted in different construction business systems?
- Can different interpretations of the client role be related to different levels of performance by the supply chain?

4.7. Concluding comments

This chapter has set out the client role, as formulated by the ICCF, and considered recent developments in the client practice, which have the effect of driving innovation through the project team. The ability to innovate in these ways, and to promote consequent change in others, clearly represents a key client capability. Initiatives such as that of the SCCF, aimed at developing a cadre of client professionals, are therefore to be encouraged, as an important means of improving construction performance overall. But significant issues remain to be addressed. Some have been reflected in the strategic programme of the European Construction Technology Platform.[8] Research conducted through this and other programmes can assist the development of a more informed and more professional client function, which through project and programme leadership will exert a positive influence on the performance of the supply side of construction.

Notes

[1] Further information about the membership and activities of the Forum, and summaries of the presentations at its meetings, may be found on www.cibworld.nl/iccf.

[2] Presentations on reform initiatives in different countries were made at the first Revaluing Construction conference (Manchester, 2003) (www.revaluing-construction.com) and the second conference (Rotterdam, 2005) (www.rc2005.org).

[3] A general review of 'value' concepts in the built environment is provided in 'Be Valuable – a guide to creating value in the built environment' published in 2006 by Constructing Excellence, London (ISBN 1–905033-14–1).

[4] The presentation by Francoise Szigetti at the first ICCF meeting on 'Performance-based Approach to Capital Assets' includes references to ISO and ASTM documents.

[5] Pepi Silinga, CEO of the Coega port development near Port Elizabeth discussed how social considerations had been taken into account in the Coega development in his presentation at the second ICCF meeting.

[6] An example of a pre-qualification process, which seeks to investigate collaborative skills is that employed by the Queensland government in Australia – see www.build.qld.gov.au/industry.

[7] Henrik Bang, Director of the Danish Construction Clients Association, emphasised this in his presentation at the first ICCF meeting.

[8] The European Construction Technology Platform (www.ectp.org) brings together a wide range of construction interests to provide a focus for construction-related research undertaken in the seventh Framework Programme of the European Commission.

References

Agency for Enterprise and Housing (EBST) (2004) *Vejledning I Partnering ('Guidelines for Partnering')*. Copenhagen, EBST.

Business Vantage (2004a) *Equal partners—customer and supplier alignment in public sector construction*. Available at: www.businessvantage.co.uk [Accessed March 2007].

Business Vantage (2004b) *Equal partners—customer and supplier alignment in private sector construction*. Available at: www.businessvantage.co.uk [Accessed March 2007].

Courtney, R.G., Rigby, J., Winch, G.M. and Bleda, M. (2006) *Study of impact of certain aspects of Community policy on the construction sector—final report*. Available at: http://ec.europa.eu/enterprise/construction/compet/policies_en.htm [Accessed March 2007].

European Commission (2004a) *Co-ordination of procedures for the award of public supply, public service and public works contracts*, Directive 2004/17/EC Official Journal 30 April 2004.

European Commission (2004b) *Co-ordination of Entities in the Water, Energy and Transport Sectors*, Directive 2004/18/EC Official Journal 30 April 2004.

Office of Government Commerce (OGC) (2003) *Procurement and Contract Strategies, Procurement Guide 6*. London, OGC.

Spencer, N.C. and Winch, G.M. (2002) *How Buildings Add Value for Clients*. London, Thomas Telford.

Swedish Construction Clients Forum (SCCF) (2006) *The Role and Mission of the Construction Client*, Stockholm.

Winch, G.M. (2000) Construction business systems in the European Union. *Building Research and Information* 28 (2): 88–97.

5 Challenging the illusion of the all powerful clients' role in driving innovation

Martin Sexton, Carl Abbott and Shu-Ling Lu

5.1. Introduction

The construction client is a significant arbitrator of value of the products and services delivered by the project team and its upstream supply chain partners. The clients' traditional view of value is one of 'exchange' value – the amount they are willing to pay, representing revenue to the project team and the construction sector as a whole. This conceptualisation is expanding to more explicit 'use' value, defined as the subjective valuation of the consumption benefits accrued from the delivered product and service; and, 'process' value, the clients' experience of the process of the product or service delivery.

Clients are increasingly demanding better performance from the project team in all of these value creation dimensions. The construction industry is thus being challenged to bring about successful innovation to create new levels of value for the client. The focus of this chapter is to challenge the prevailing mantra that clients must drive innovation, and that clients should capture all the benefits of innovation. This is a dangerous illusion, which denies the reality that different stakeholders 'drive' innovation in different ways and benefit from innovation in a similarly differentiated fashion. If clients assume the role of innovation champion on all occasions, the motivation of other stakeholders to participate in the innovation process will wane – to the detriment of all. The challenge is for clients to adopt an appropriate role in innovation processes. This chapter provides an overview of the definitional debate surrounding the concept of innovation. From this, three archetypical client roles in the innovation process are delineated: dominant, balanced co-production and passive. These roles offer a contingency portfolio of roles and responses to meet the particular characteristics and needs of a given situation. The chapter moves on to using the case of performance-based building (PBB) as a vehicle to demonstrate the contingency approach advocated. Finally, conclusions are drawn.

5.2. Definition of innovation

In the general management literature, innovation tends to be generically defined as being something new, which is implemented by a firm in some way. For example, innovation is defined as 'generation, acceptance and implementation of new ideas,

processes, products or services' (Thompson, 1965, p. 36) or the 'successful implementation of creative ideas within an organisation' (Amabile *et al.*, 1996, p. 25). The construction literature is consistent with the general literature, with, for example, innovation
being defined 'as the actual use of a non-trivial change and improvement in a process,
product or system that is novel to the institution developing the change' (Slaughter,
1998, p. 226) or 'the process of bringing in new methods and ideas or making changes'
(Atkin and Pothecary, 1994, p. 55).

What is interesting is that such definitions of innovation are 'value' neutral – namely,
they do not explicitly state that innovation should add value in some way to a particular stakeholder group (say, the firm or the client) if it is to be deemed successful. This
'value' neutrality highlights a dominant assumption in the literature; namely, that all
innovation is beneficial. Kimberly (1981, p. 84–85) brings attention to this by noting
that 'innovation tends to be viewed in unreflective positive terms ... [and that] ... for
the most part, researchers have assumed that innovation is good'. This assumption
hampers an appreciation that innovation is associated with uncertainty and the risk
of failure. Capaldo *et al.* (1997), for example, stress that innovation does not lead me-
chanically to improved performance – on the contrary, the decision to innovate may
even strongly jeopardise the firm. The risk of such jeopardy leads to the 'innovator's
dilemma' (Christensen, 1997): under which conditions firms should stick to what they
already do and in which situation they should initiate innovation activity.

In the general literature there are hardening pockets calling for the term innovation
to accommodate the explicit benefit, which must flow if innovation is to be considered
successful. This is emphasised in the observation that 'innovation consists of the generation of a new idea and its implementation into a new product, process or service,
leading to the dynamic growth of the national economy and the increase of employment
as well as the creation of pure profit for the innovative business enterprise' (Urable,
1988, p. 3). Similarly, in the construction literature, for example, it is suggested that
successful innovation is 'the effective generation and implementation of a new idea,
which enhances overall organisational performance' (Sexton and Barrett, 2003, p. 626).

In summary, innovation activity must, if it is to be sustainable, benefit all stakeholders in the construction process – not just the client! This understanding is manifest
in Barrett (2005, p. 1) in the assertion that for construction to be revalued, there is a
need for 'the maximisation of value jointly created by stakeholders to construction
and the equitable distribution of the resulting rewards'. This multi-stakeholder view
of innovation cautions against the prescription that clients should drive (and benefit
from) all innovation. Rather, it clarifies the need that the stakeholder driving innovation must be appropriate for a given situation. From this context, the following
section will offer a differentiated, contingency-based view of clients' roles in driving
innovation.

5.3. Clients' role in the innovation process

The clients' role in the innovation process can be located along a continuum: a dominant
role – where clients drive innovation; a balanced co-production role – where the project
team and the client jointly drive innovation; and, a passive role – where the project
team drive innovation. These different roles will be described in turn, drawing upon
the general and construction-specific innovation literature.

5.3.1. Dominant client role

The idea of the dominant client role in the innovation process is typified in the concept of the 'lead user'. von Hippel (1986) demonstrates that lead users in a variety of industries (particularly the manufacturing sector) play a significant role in the innovation process, especially in the invention and early innovation phases. Lead users were discerned to have two principal characteristics: they address needs that will be widespread in the market place, but they face them months or years before the majority of those in the market place encounter them; and, their position in the market place allows them to benefit substantially from obtaining a solution to their needs. These lead-user attributes combine to drive innovation at a sector level as well as for the lead users' particular needs.

In a construction context, the lead user role can be witnessed, for example, in the way the British Airports Authority (BAA) in the UK took a leading role in driving the development and implementation of partnering, which became the bedrock of the Egan agenda (Miozzo and Ivory, 2000, p. 518), for instance, captured the underpinning lead user client role, in the comment that the BAA's procurement programme 'has been a test case for many of these [Egan agenda] ideas', and that, in its wake, 'other large client organisations have also adopted versions of partnering in their procurement, including Sainsbury, Tesco, Rover, Water Companies and the UK government' This guiding coalition of lead users have encouraged (indeed, demanded) an ongoing shift in the sector to more collaborative ways of working.

While lead users are clearly important in the innovation process, not all users are equal in value with respect to the capacity and capability to lead sector level innovation. The majority of clients involved in specific projects and partnering arrangements engage and influence innovation through co-production processes at a project level with a particular project team.

5.3.2. Balanced co-production role

The general management literature has long documented the role of co-production in the innovation process. Lundvall (1988), for example, argues that most innovations are developed during collaboration between users and producers. Recent work on complex products and systems has produced alternative models of the organisation of innovation activity that include high levels of client involvement in the design of products and temporary networks of production and delivery (e.g. DeFillippi and Arthur, 1998).

The co-production of innovation within the construction domain is evident in the briefing literature with, for example, Barrett and Stanley (1999) seeing it as a very interactive process between the client and project team focused around empowering the client, managing the project dynamics, appropriate user involvement, appropriate visualisation techniques and appropriate team building. The joint production of innovation is also demonstrated in professional service activity, with (Lu and Sexton, 2006, p. 1270) stating that successful innovation is 'principally the outcome of the co-production between the knowledge worker and the client'.

The co-production approach is the 'bread and butter' of client role in the innovation process within one-off commissions and partnering ventures. The client and the project team contribute to the development of innovative goals and solutions. The dominant and balanced co-production roles are, to varying degrees, 'client-intensive', requiring inputs from the client into the design and production phase. Many clients, however, purchase the outputs of the design and production phases without any engagement in those phases. These clients play a passive role in the innovation process.

5.3.3. Passive client role

The passive client role in innovation is the 'mass market' client that consumes 'off the shelf' products and services with very little influence on their key attributes. This client involvement generally limits their role in product/service development to delivering feedback voluntarily to the producer or when requested to do so by market researchers. This feedback can be a valuable source of information in the improvement of existing products and services. The drawback of this type of feedback, however, is that it is unlikely to lead to radically new products or services. Clients' or potential clients' mindsets on which the feedback is based is anchored to present products and services, preventing users from thinking about their attributes from a new perspective – an effect called 'functional fixedness' (Luthje *et al.*, 2002, p. 8).

In a construction context, this is illustrated by the consumers of speculative housing and office space. The clients' role in the innovation process is very much at arm's length, based predominantly on making complaints to the developer/owner once the facilities has been bought and defects are detected. This feedback, in theory, will influence future design solutions. The feedback loop, however, is often incomplete and slow.

Three client roles have been identified, each operating within different principal loci: dominant client role – sector and project innovation; co-production client role – project innovation; and, passive client role – provision of feedback for future product innovation. The following section illustrates these loci of client intervention using PBB as a case study.

5.4. The role of the client in performance-based building innovation

PBB is intertwined with the present interest in, and momentum towards, PBB codes and standards. The key driver for this trend is the view that traditional prescriptive approaches act as a barrier to innovation. Two schools of thought regarding the relationship between PBB and innovation have been identified (Sexton *et al.*, 2005). The 'content' school of thought has advocated that PBB is the innovation in itself, and that PBB approaches replace traditional prescriptive approaches with a new paradigm. In contrast, the 'context' school of thought has argued that PBB provides the enabling environment to stimulate a raft of innovation activity, which may include prescriptive, as well as performance-based, elements. For the purposes of this chapter, the focus will be limited to the context school.

The 'context' school of thought views PBB as a guiding framework, which provides a stimulating, supportive environment that encourages innovation, be it traditional, prescriptive or PBB codes and regulations, or a combination of the two, to provide buildings that 'meet all the goals established by society and the client' (Averill, 1998, p. 2). This school of thought is grounded on two key assumptions. First, it adopts a contingency premise that the PBB approach is not universally applicable to all projects

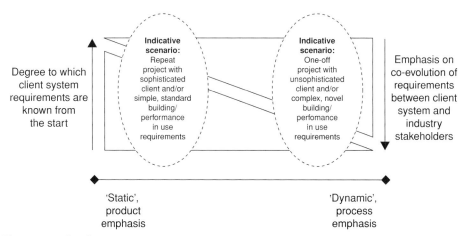

Figure 5.1   Product process innovation continuum.

in all circumstances. Rather, a contingent view of PBB is promoted, which appreciates projects with different characteristics and circumstances require different mixes of PBB and/or prescriptive elements. This argument is captured in Figure 5.1 which suggests, on the left hand side of the diagram, that where the client system requirements are known from the start, a 'product' view of innovation is appropriate and that this is best supported by client engagement towards the *passive* end of the client role in the innovation process continuum described in the previous section. Indicative scenarios of this position would be repeat projects with a sophisticated client and/or simple, standard performance requirements. In contrast, where there is a need for a co-evolution of requirements between the client system and the project team, as depicted on the right hand side of the diagram, a 'process' view of innovation is appropriate and best sustained through client involvement in the *balanced co-production* portion of the client role in the innovation continuum. Indicative scenarios would be a one-off project with an unsophisticated client and/or complex, novel performance requirements.

Second, the 'context' school emphasises a social systems perspective which argues that innovation is determined by the nature and intensity of interactions, interconnectedness and synergies from a wide spectrum of agents which gravitate around a project setting. Lead users can play a *dominant* client role in developing and/or influencing inter-organisational networks that promote and facilitate the development and exchange of knowledge and resources that are needed to encourage learning and innovation in participating companies.

5.5. Conclusions

At the beginning of this chapter, the blanket assertion that clients should drive innovation was challenged. A contingency approach to the role of client in the innovation processes has been set out, identifying three archetypical roles: dominant, balanced co-production and passive. The different roles reflect and progress different configurations of client, project team and project characteristics and needs. The key criterion for selecting the client role is *appropriateness*; i.e. clients appropriately driving innovation. Further, the flow of exchange, use and process value created through the innovation process is, echoing Barrett (2005), maximised and equitably distributed among stakeholders.

References

Amabile, T.M., Conti, R., Coon, H., Lazenby, J. and Herron, M. (1996) Assessing the work environment for creativity. *Academy of Management Journal* **39**: 1154–1184.

Atkin, B. and Pothecary, E. (1994) *Building Futures: A Report on the Future Organisation of the Building Process*. Reading, England, University of Reading.

Averill, J.D. (1998) Performance-based codes: economics, documentation and design. *MSc Thesis*, Worcester Polytechnic Institute, USA.

Barrett, P. (2005) *Revaluing Construction: A Global CIB Agenda*. CIB Report 305, CIB, Rotterdam.

Barrett, P. and Stanley, C. (1999) *Better Construction Briefing*. Oxford, UK, Blackwell Science.

Capaldo, G., Corti, E. and Greco, O. (1997) A coordinated network of different actors to offer innovation services to develop local SMEs inside areas with a delay of development. *Proceedings of the ERSA Conference*, 26–29 August, Rome.

Christensen, C.M. (1997) *The Innovator's Dilemma: When New Technology Cause Great Firms to Fail*. Boston, Harvard Business School Press.

DeFillippi, R. and Arthur, M. (1998) Paradox in project-based enterprise: the case of film making. *California Management Review* **40** (2): 125–139.

Kimberly, J.R. (1981) Managerial innovation. In: *Handbook of Organizational Design—Adapting Organizations to Their Environments* (eds P.C. Nystrom and W.H. Starbuck), Vol. 1. New York, Oxford University Press, pp. 84–104.

Lu, S. and Sexton, M. (2006) Innovation in small construction knowledge-intensive professional service firms: a case study of an architectural practice. *Construction Management and Economics* **24** (12): 1269–1282.

Lundvall, B.A. (1988) Innovation as an interactive process—From user-producer interaction to national systems of innovation. In: *Technical Change and Economic Theory* (eds G. Dosi, C. Freeman, R. Nelson, G. Silverberg and L. Soete). London, Pinter.

Luthje, C., Herstatt, C. and von Hippel, E. (2002) *The Dominant Role of 'Local' Information in User Innovation: The Case of Mountain Biking*. MIT Sloan School Working Paper, MIT.

Miozzo, M. and Ivory, C. (2000) Restructuring in the British construction industry: implications of recent changes in project management and technology. *Technology Analysis & Strategic Management* **12** (4): 513–531.

Sexton, M. and Barrett, P. (2003) Appropriate innovation in small construction firms. *Construction Management and Economics* **21**: 623–633.

Sexton, M., Barrett, P. and Lee, A. (2005) The relationship between performance-based building and innovation: an evolutionary approach. *Proceedings of the 11th Joint CIB International Symposium: Combing Forces—Advancing Facilities Management and Construction Through Innovation*, 13–16 June, Helsinki, Finland.

Slaughter, S.E. (1998) Models of construction innovation. *Journal of Construction Engineering and Management* **124** (3): 226–231.

Thompson, V.A. (1965) Bureaucracy and innovation. *Administrative Science Quarterly* **5**: 1–120.

Urable, K. (1988) *Innovation and Management*. New York, Walter de Gruyter.

von Hippel, E. (1986) Lead users: a source of novel product concepts. *Management Science* **32** (7): 791–805.

6 Reifying the client in construction management research? Alternative perspectives on a complex construct

Mike Bresnen

6.1. Introduction

Many recent management initiatives in construction place the client much more centre stage in the construction procurement, design and management process. Partnering, for example, is often based on clients taking a more proactive approach to the management of their projects than hitherto (Barlow *et al.*, 1997). However, despite the greater importance attached to the role, conceptualisations of 'the client' tend to remain firmly rooted in a view of organisation that emphasises their unitary nature – downplaying the impact of diverse stakeholders whose interests and interactions crucially influence project decision-making (Newcombe, 2003).

This chapter takes as its point of departure the idea that talking about 'the client' operates as a form of reification. That is, it attributes human-like qualities to abstract objects (in this case, complex client organisations) while, at the same time, objectifying social relations (so that clients 'act' as monolithic entities). Consequently, there is still a tendency to assume unitary behaviour on the part of clients and, with some exceptions (e.g. Walker and Newcombe, 2000), to downplay issues of power, conflict and control within client organisations and between clients and other participants in the project management process.

However, the chapter goes further, in proposing that stakeholder analysis that takes a more pluralistic approach to understanding diversity and conflict within (client) organisations can only go so far. In particular, such an approach obscures built-in cultural assumptions that enable the mobilisation of bias in decision-making and the operation of more subtle forms of control based on institutionalised normative systems (Lukes, 1974). It also obscures the potentially totalising effect of systems of surveillance that create 'normalised' behaviour within complex organisations (Foucault, 1980). Drawing upon alternative perspectives from organisational theory (Bresnen *et al.*, 2005) and also from recent empirical work on clients and partnering (Bresnen and Marshall, 2000), this chapter assesses the implications for innovation of taking a more critically informed approach to understanding the nature of construction industry clients.

6.2. The construction client in theory and practice

'Meeting the client's needs' has become something of a leitmotif in recent years in industry and government reports that have recommended widespread change in the way the industry operates (e.g. Latham, 1994; Egan, 1998). Clients are no longer seen as the relatively passive actors in the project management process they once were and are now considered highly proactive in shaping not only the forms of delivery system that are used, but also the processes and outcomes that they produce (Briscoe *et al.*, 2004). Research on partnering, for example, has highlighted how important clients are to establishing, developing and sustaining more collaborative relationships with contractors (Barlow *et al.*, 1997; Bresnen and Marshall, 2000). A wide range of managerial initiatives – from lean construction to benchmarking, continuous improvement and the like – are based on the premise that introducing new management ideas (drawn from manufacturing industry) can promote better project integration and overall performance to the ultimate benefit of the client (Gann, 1996). As such, there is a strong sense in which the 'cult' of the customer established in manufacturing and service industries from the 1980s onwards has become firmly entrenched within the construction sector (Ivory, 2005).

Leaving aside for the moment the difficult questions of how successful such initiatives are and whose interests they ultimately reflect and serve (e.g. Green, 1998), there is an obvious problem in the way in which the client is represented in such discourses. More often than not, 'the client' is presented in an almost anthropomorphic way – taking on individual human attributes, despite the fact that clients are often sizeable private or public sector organisations (e.g. Bresnen and Haslam, 1991). This is also in spite of importance of the 'client's representative' – the person assigned to act on behalf of the client and the supposed embodiment of the client's needs. Even where there is clear recognition that the client is an organisation, there is too often a tendency to talk of 'the client' and 'their needs' as if they constitute a single organisational entity that acts coherently and consistently towards others in the project. Yet, this does not adequately reflect the more complex, pluralistic nature of clients that has been recognised for some considerable time (Bryant *et al.*, 1969).

The result is the creation of a highly reified view of 'the client'. Reification refers to the tendency to attribute human-like qualities to abstract objects (in this case, complex client organisations) while, at the same time, implying the objectification of social relations (where clients 'act' as unitary, monolithic entities) (Petrovic, 1983). Consequently, the reification of clients involves the reduction of the complex organisations that constitute clients in practice to a much simpler abstraction. In this simplified view of organisations, 'the client' not only possesses hard, objective qualities (the organisation is 'real'), but also 'speaks' with one voice, pursues one set of values and goals and engages in purposeful action. Furthermore, the client forms part of a larger community of clients, who have more or less common interests and needs (that the industry is then expected to meet).

Research on construction organisations in general, and on client organisations in particular, has, of course, taken a more penetrating look at both the nature of this community of clients and at the nature of client organisations themselves. First, there is the clear recognition that clients differ in many obvious and important ways (public/private, scale, experience, etc.) that affect the extent to which they are able to define, articulate and pursue their interests and goals on projects. Type and level of experience, for example, are important variables that may affect the way in which clients approach many aspects of project procurement and management (Bresnen and Haslam, 1991).

Second, there is a well-established literature that takes a more pluralistic approach to the study of organisations in construction and which recognises the importance of the internal political working of organisations (Walker and Newcombe, 2000). Recent work, in particular, has applied the logic and methods of stakeholder analysis in order to unpack the diverse interests and sources of influence that are brought to bear on project decision-making (e.g. Newcombe, 2003). In this approach, attention is directed towards the motivation (aims) and ability (power) of groups to pursue their interests and the coherence of the client as a unitary entity is not necessarily taken for granted (Newcombe, 2003). Instead, emphasis is placed on exploring project decision-making in a context in which diverse groups that cut across organisational boundaries need to collaborate to achieve common project goals – despite their divergent interests (Cherns and Bryant, 1984).

However, as will be seen, such approaches, although extremely valuable, only go so far, in that they tend to leave unexplored and unquestioned the more deeply embedded values, norms and practices within which such interactions play out. In order to explore these deeper influences, it is necessary to look further into alternative perspectives on organisation that tend to receive comparatively less attention in the construction management literature.

6.3. Alternative perspectives on (client) organisations

Elsewhere, it has been argued that relatively few attempts have been made within the construction management field to expand beyond the straitjacket of positivism and to draw upon the wider repertoire of critical social science thinking that constitutes organisation studies (Bresnen *et al.*, 2005). Although critical perspectives can vary enormously – encompassing modernist and post-modernist thinking on organisation (Cooper and Burrell, 1988) – they do share a number of common concerns (Bresnen *et al.*, 2005). First, an acknowledgement of the importance of power and politics in understanding organisational behaviour and management – not simply as 'variables' to be explored, but as a way of conceptualising relations within and between organisations. Second, a commitment to questioning and deconstructing commonly held notions, icons and images of organisation (e.g. 'the client'). Third, a concern with epistemology and a questioning of the knowledge base of claims made about what is 'objectively true' about organisations and their management. Fourth, a commitment to an investigative stance that questions taken-for-granted values, assumptions and meanings.

A post-modernist perspective might, for example, go even so far as to ask why we tend to conceive of 'the organisation' as a structured, bounded phenomenon that exists in a more or less ordered and given social context? As opposed to seeing organising as a skilled social accomplishment, albeit inevitably transient and fragile, that involves the active bringing together (and fortuitous coming together) of flows of action and information in what is otherwise a complex web of changing social, economic and political relations (Tsoukas and Chia, 2002). Which of these two ontological positions in fact resonates better with what we know about the problems of organising and managing projects within the intricate network of intra- and inter-organisational relations in which they are embedded (Sydow *et al.*, 2004)?

To apply the range of critical perspectives alluded to above is much too ambitious a project. However, a start can be made by separating out and applying some ideas about power that are well established as part of this more critical perspective on organisation, but which go beyond the pluralism implied by stakeholder analysis. The following section attempts to do this, elaborating a framework for exploring the deeper processes of structuring and enculturation that relate to construction industry clients, using illustrative quotes drawn from recent research by the author (Bresnen and Marshall, 2000).

6.4. A critical view of power and the client

The quotations used in this section are all taken from an interview with the property development director of a large UK hotel chain, who was questioned about a new hotel project and his company's approach to partnering. The first simply illustrates how, even within the relatively narrow confines of the client's property development directorate, it becomes difficult to presume that clear boundary lines can be drawn in talking about 'the client':

> [Referring to organisation chart:] That's me and this is the team. Within this team, we include our core consultant team and our core building team and they're alongside our employees. . . . We've almost got a virtual company here and I don't really know how to describe this organisational structure, but it's sort of a group of people with common objectives. They aren't all employees.

Beyond the core property development team, it was clear, however, that there were major differences in perspective and interests that needed to be considered. Referring, for example, to reactions to the introduction of partnering:

> As you move from one-off competitive tendering as a procurement route to partnership sourcing, the employees within the business start to feel that they're turkeys waiting for Christmas, because they can see that, [as] they are working with a smaller number of people and getting closer and closer to them, you don't need the same administrative set-up as you previously had with an adversarial arrangement and they start to sabotage it.

Pluralist approaches to organisation have long recognised the importance of diversity in interests, goals and power. Work that takes such an approach – including stakeholder analysis – reflects a long-established mainstream management tradition in which power relations amongst divergent interests play a very important part (Hardy, 1994). However, such work has tended to stay within a positivist tradition, exploring power-dependency relations within and between management groups and their effect on conflict. It has not gone further to explore the deeper structuring of power within social organisations which shapes such conflict by determining the extent to which action might be considered 'legitimate' or 'illegitimate' (Hardy, 1994).

Lukes (1974) was among the first to highlight how such orthodox treatments tend to encompass only two 'dimensions' of power. The first and most obvious dimension concerns the analysis of decision-making in conditions of overt conflict, where power resources (financial, expertise, etc.) are mobilised and used to pursue vested interests (often leading to negotiated solutions). The second dimension of power, however, recognises that power can also be used to *prevent* decision-making and stop conflict becoming overt – through the mobilisation of built-in structural biases that favour the sovereign power of the hegemonic group. So, for example, agendas, terms of reference and voting systems can all be used by those in positions of dominance to suppress conflict by preventing participation in decision-making and avoiding the need to accommodate diverse interests.

Together these two dimensions amount to what Hardy (1994) describes as 'instrumental power'. Lukes (1974), however, went further to argue that the use of power was often more subtle than this and involved the manipulation of language, values and norms in ways that could prevent conflict from occurring in the first place – thereby directly serving the interests of the dominant group. This third dimension of power places emphasis on what action is considered 'legitimate' or 'illegitimate' and is of crucial importance in understanding how power works at a deeper structural/cultural level – by effectively *preventing conflict from emerging* through the institutionalisation of values and norms (Hardy and Clegg, 1999).

Hardy (1994) describes this form of power as 'symbolic power', which works through the 'management of meaning'. So, for example, language can be used 'as a catalyst to mobilise support, or as a device to cloud issues and quiet opposition' (Hardy, 1994: 226). As testimony to this, it is not too difficult to see the attempts made by management to move beyond bureaucratic forms of control and to 'manage culture' through appealing to wider organisational values and norms (Willmott, 1993). More specifically, and in relation to the use of symbolic power within clients, consider the following account of the strategies used to gain support for the roll out of partnering and how an association between partnering and perceptions of effective performance is consciously created:

> You have to start, I think, with creating unhappiness with the status quo at the highest possible level of the organisation. The best way of doing that is to demonstrate that other people can do what you're doing better than you are . . . so that you can start to challenge and say: 'look, we've been building like this since 1740. Other people adopting these forms of construction are doing it much quicker. The reason they've done this is because they've got close working relationships with people, whereas we haven't' . . . . So you get them on the hook with that . . . you start to create some unhappiness at that level . . . and then you reel it in a bit at a time.

The concept of symbolic power is important in highlighting the more unobtrusive and yet profound influence of power. It also crucially emphasises the more distributed nature of power – in the language, rituals and myths used by subcultures and counter-cultures, for example. Such symbolic power resources may, in turn, form part of the arsenal that project stakeholders are able to use in collective decision-making (though they would still be subject to whatever 'rules of engagement' had been negotiated). Yet, as with instrumental power, the assumption is that there is some basic level of agency and intention on the part of a sovereign power that is somehow able to consciously use symbolism (e.g. articulation of core values) to manipulate meaning and thus influence others (Hardy, 1994).

Such sovereign power 'over' others (Marshall, 2006) may well be important, but it is not the whole story. Further developments in the conceptualisation of power have drawn upon post-modernism to draw out a fourth and even more subtle, dimension. Drawing on (Foucault's 1977, 1980) work, Hardy (1994) defines this fourth dimension as the power of the system, relating it to 'unconscious acceptance of the values, traditions, cultures and structures of a given institution or society' (Hardy, 1994: 230). In this perspective on power, conflict may not even be perceived or felt as it resides in 'normalised' modes of behaviour and concepts of self-identity which are internalised and which tend to be reflected in, and reinforced by, everyday behaviour. Consider, for example, the following depiction of the client as 'good citizen' in accommodating diverse needs and interests in planning for a new hotel:

> Planners have requirements that have to be accommodated. . . . The local community's requirements have to be accommodated. *We are anxious to be good neighbours,* so we try to take account of the local community's needs. (Emphasis added.)

According to this perspective, power and knowledge are inextricably intertwined: knowledge is power and existing configurations of power/knowledge reflect and reinforce deeply held internalised values and norms (Hardy and Clegg, 1999). The emphasis here is upon disciplinary power – both in the sense of control of the individual (through surveillance to ensure compliance through self-control); and, in the sense of how the definition of what constitutes a body of knowledge operates in itself as a system of power (Hodgson, 2002). So, for example, one can see in the practices developed to refine and advance knowledge at the local level, how the micro-politics of power operate to shape and reinforce expected (and uncontested) self-governing behaviour.

Consider, for example, the following account of the staff development of hotel property managers, which illustrates the disciplinary effects of mechanisms (namely, focus groups) that make property managers' performance and attitudes subject to very direct forms of surveillance:

> We would involve [them in] focus groups to try and instil our learning. . . . He's almost like a premises manager but it's very useful to be growing those individuals in our learning. Progressively, we want these hotel-based property managers to take more and more responsibility for their own hotels . . . and so we're trying to grow these people into being something more than having a screwdriver in the top pocket – more like a manager who can have a five-year plan for his hotel.

Similarly, in the following quote, we see a system in operation that depends for its effect upon a collective willingness by the group to challenge its own efforts and achievements:

> Michael, who is our special project manager, is the "keeper of [design and cost] models". He looks after them, he develops them, he works on them. He is constantly looking to refine, enhance and improve them and make them better – always looking at innovative ways of doing things, with the core consultancy team. So they are expected as a group, whether we're building anything or not, to constantly be challenging that model and trying to drive it to be more efficient and effective and innovative.

There is considerable debate within the literature about the totalising effects of surveillance as a means of control and therefore about the opportunities for resistance and emancipation (Hardy and Clegg, 1999, pp. 379–381). However, there is also recognition that 'the power of the system' can be harnessed in the interests of those less powerful – as much as by any elite group whose interests it might more closely represent. Significantly, for example, many of the above quotes highlight the limits to power and influence felt by the respondent vis-à-vis his own firm – despite his position of power as property development director.

6.5. Implications for clients driving innovation

The above discussion suggests a number of important implications for understanding client organisations and their role in innovation. First, the importance of 'instrumental power' in organisations draws attention to the obvious difficulty in assuming that 'clients' have a coherent and uncontested agenda when it comes to innovation or any other organisational process. Not only may there be differences of interest and perspective within the organisation's dominant coalition, but these differences may also become more amplified the more one moves out across the (horizontally and vertically differentiated) client organisation. Such differences may, of course, be moderated where there are stronger over-riding structural mechanisms and/or unifying organisational values and norms.

Whatever the situation, any differences are likely to play out in a number of ways where decision-making is localised, power is distributed and influence is based on a range of hierarchical and non-hierarchical sources (expertise, etc.) (cf. Sapsed and Salter, 2004). So, for example, attempts to introduce new management initiatives, such as partnering, may be inhibited by dispersed management practices and routines (Bresnen *et al.*, 2004). Moreover, differences in aims or perspective may heighten the organisation's vulnerability in interaction with third parties both internally (e.g. other departments) and externally (e.g. contractors, designers) – creating greater uncertainty and caution in interaction, or providing other parties with the chance to pursue their own interests (Bresnen, 1990).

Second, the importance of 'symbolic power' in organisations draws attention to the impact of discourse upon clients' actions and perceptions of 'client needs'. It may be, for example, that a dominant organisational discourse emerges regarding innovation and the need for change (or perhaps several competing discourses). The discursive resources brought to bear (language, value statements, etc.) may reinforce particular ways of articulating needs, creating agendas for change, privileging certain performance criteria and the like that have subtle, but very powerful (and sometimes unintended) effects on innovation and other organisational processes. So, for example, the clarion call to respond to the needs of clients is not necessarily consistent with innovation (Ivory, 2005); nor is the shorter-term orientation within the industry towards immediate project performance (Winch, 1998). Furthermore, dominant discourses within client organisations are bound to be influenced by discourse at a wider, institutional level across the industry (Green and May, 2005). Through these, common conceptions of effective performance, innovation, client 'best practice' and the like are likely to be spread by isomorphic pressures to emulate others' successes – notwithstanding whether or not such practices are suited to 'the needs' of particular clients.

Finally, the importance of 'disciplinary power' in organisations draws attention to patterns of power/knowledge within the organisation and conceptions of identity as they shape and reinforce particular ways of thinking and acting. In contrast with 'stakeholder analysis', the point being made here is that notions of in-group and out-group membership, as well as of what constitutes normalised ways of working within and between groups, are highly fluid. This makes it difficult to establish clear boundaries around political entities; or to presume that such interests and perspectives can be represented statically or without recourse to the broader structures of power in which they are embedded and in whose reproduction through practice they are actively engaged (cf. Giddens, 1990). In other words, stakeholders do not interact upon a relatively fixed and level playing field, but according to the arenas and 'rules of engagement' established historically and which are reconstituted through continuing interaction and negotiation amongst interested parties.

With regard to clients and innovation, a focus on disciplinary power points again to the importance of discourse as the means through which conceptions of client interests and needs (and therefore of those supposedly 'best placed' to articulate those needs) are shaped and moulded. So, for example, particular ways of talking about and promoting innovation (e.g. by encouraging contractor participation in design processes or by privileging technical accomplishments) come to represent particular configurations of power/knowledge within the organisation around which behaviour is 'normalised'. These, in turn, tend to reinforce prevailing patterns of influence within the client organisation (leading to the hegemony of groups who advocate those positions) and also reflect the impact of wider, institutional-level discourse within the sector (thus providing the basis for legitimacy of action). Put another way, the creation and diffusion of innovation and 'best practice' amongst and within client organisations are highly social and political processes. Power and knowledge are closely interconnected and what constitutes 'innovation' and 'best practice' is constructed, reinforced and reproduced (and perhaps ultimately changed) through social interaction and joint practice.

References

Barlow, J., Cohen, M., Jashapara, A. and Simpson, Y. (1997) *Towards Positive Partnering.* Bristol, The Policy Press.

Bresnen, M. (1990) *Organizing Construction: Project Organization and Matrix Management.* London, Routledge.

Bresnen, M., Goussevskaia, A. and Swan, J. (2004) Embedding new management knowledge in project-based organizations. *Organization Studies* **25** (9): 1535–1555.

Bresnen, M., Goussevskaia, A. and Swan, J. (2005) Managing projects as complex social settings. *Building Research and Information* **33** (6): 487–493.

Bresnen, M. and Haslam, C. (1991) Construction industry clients: a survey of their attributes and project management practices. *Construction Management and Economics* **9** (4): 327–342.

Bresnen, M. and Marshall, N. (2000) Building partnerships: case studies of client-contractor collaboration in the UK construction industry. *Construction Management and Economics* **18** (7): 819–832.

Briscoe, G.H., Dainty, A.R.J., Millett, S.J. and Neale, R.H. (2004) Client-led strategies for construction supply chain improvement. *Construction Management and Economics* **22** (February): 193–201.

Bryant, D.T., Mackenzie, M.R. and Amos, W. (1969) *The Role of the Client in Building.* London, Tavistock.

Cherns, A.B. and Bryant, D.T. (1984) Studying the client's role in construction management. *Construction Management and Economics* **2**: 177–184.

Cooper, R. and Burrell, G. (1988) Modernism, postmodernism and organizational analysis: an introduction. *Organization Studies* **9** (2): 91–112.

Egan, J. (1998) *Rethinking Construction*. London, DETR.

Foucault, M. (1977) *Discipline and Punish: The Birth of the Prison*. Harmondsworth, Penguin.

Foucault, M. (1980) The eye of power. In: *Power/Knowledge: Selected Interviews and Other Writings* (ed. C. Gordon), 1972—1977. Hemel Hempstead, Harvester Wheatsheaf, pp. 146–165.

Gann, D.M. (1996) Construction as a manufacturing process? Similarities and differences between industrialised housing and car production in Japan. *Construction Management and Economics* **14**: 437–450.

Giddens, A. (1990) *The Consequences of Modernity*. Stanford, CA, Stanford University Press.

Green, S.D. (1998) The technocratic totalitarianism of construction process improvement: a critical perspective. *Engineering, Construction and Architectural Management* **5** (4): 376–386.

Green, S.D. and May, S.C. (2005) Lean construction: arenas of enactment, models of diffusion and the meaning of 'leanness'. *Building Research and Information* **33** (6): 498–511.

Hardy, C. (1994) Power and politics in action. In: *Managing Strategic Action* (ed. C. Hardy). London, Sage, pp. 220–237.

Hardy, C. and Clegg, S.R. (1999) Some dare call it power. In: *Studying Organization: Theory and Method* (eds C. Hardy and S.R. Clegg). London, Sage, pp. 368–387.

Hodgson, D.E. (2002) Disciplining the professional: the case of project management. *Journal of Management Studies* **39** (6): 803–821.

Ivory, C. (2005) The cult of customer responsiveness: is design innovation the price of a client-focused construction industry? *Construction Management and Economics* **23** (October) 861–870.

Latham, M. (1994) *Constructing the Team*. London, HMSO.

Lukes, S. (1974) *Power: A Radical View*. London, Macmillan.

Marshall, N. (2006) Understanding power in project settings. In: *Making Projects Critical* (eds D. Hodgson and S. Cicmil). Basingstoke, Palgrave Macmillan, pp. 207–231.

Newcombe, R. (2003) From client to project stakeholders: a stakeholder mapping approach. *Construction Management and Economics* **21** (December): 841–848.

Petrovic, G. (1983) Reification. In: *A Dictionary of Marxist Thought* (eds T. Bottomore, L. Harris, V.G. Kiernan and R. Miliband). Cambridge, MA, Harvard University Press, pp. 411–413.

Sapsed, J. and Salter, A. (2004) Postcards from the edge: local communities, global programs and boundary objects. *Organization Studies* **25** (9): 1515–1534.

Sydow, J., Lindkvist, L. and DeFillippi, R. (2004) Project-based organizations, embeddedness and repositories of knowledge. *Organization Studies* **25** (9): 1475–1489. [Editorial.]

Tsoukas, H. and Chia, R. (2002) On organizational becoming: rethinking organizational change. *Organization Science* **13** (5): 567–582.

Walker, A. and Newcombe, R. (2000) The positive use of power on a major construction project. *Construction Management and Economics* **18** (1): 37–44.

Willmott, H. (1993) Strength is ignorance; slavery is freedom: managing culture in modern organizations. *Journal of Management Studies* **30** (4): 515–552.

Winch, G. M. (1998) Zephyrs of creative destruction: understanding the management of innovation in construction. *Building Research and Information* **26** (5): 268–279.

7 A proposed taxonomy for construction clients

Patricia Tzortzopoulos, Mike Kagioglou and Kenneth Treadaway

7.1. Introduction

The construction industry has been challenged to deliver better services and products to its clients. In this context, means by which appropriate value, sustainable outcomes and innovation could be achieved through a prosperous construction industry have been investigated (e.g. Ivory, 2005; Boyd and Chinyio, 2006). A better understanding of construction clients' identity and the consequent definition of a classification system is seen as essential in supporting the industry to respond to such challenges.

The development of 'a well-defined theoretical or empirical classification is a basic step in conducting any form of systematic inquiry into the phenomena under investigation' (McCarthy, 1995, p. 37); therefore it is perceived as essential. An appropriate classification for construction clients is necessary to provide clarity in terms of who they are, their needs, their likely involvement with the process and support needed. Such taxonomy may provide clarity by exposing general characteristics of clients (organisations), supporting the understanding of their interests, motivations and ability to stimulate both process and product innovation in construction. It can also assist by clarifying roles and responsibilities for specific types of client, considering the context in which they operate. Clarifying the types and interests of clients is important to allow construction professionals to take appropriate actions on each project to elucidate clients' objectives. But firstly, our present understanding of clients, and consequently the structure of a classification, needs consideration.

This chapter presents a literature synthesis focused on bringing to light different definitions and types of clients. A discussion on the ability of different clients to stimulate innovation in construction is also presented. This aims to help *determine and predict* the relationships that govern the client involvement with, and support for, innovation during the design and construction of buildings.

There are clearly diverse construction clients, each with different aspirations, motivations and willingness to innovate. However, understanding the client is not straightforward. For instance, Green (1996) argues that construction professionals need to build up good relationships with clients to be able to understand their needs, preferences and requirements. Flanagan *et al.* (1998, p. 22) emphasise that 'the generic "customer" of the past no longer exists' due to changes in the methods and types of project financing (e.g. LIFT – Local Finance Improvement Trust) (see Chapter 21 for a case study of a major capital development programme, the Manchester, Salford and Trafford Local Improvement Finance Trust), which have led to a diversification of the customer base. Newcombe (2003) further argues that issues such as the separation of ownership and occupation, the rise of the corporate client and the greater penetration in the industry by the continuing client have led to confusion about the clients' identity and their interaction with the industry. The concept of 'stakeholders' also extended the traditional understanding of the client to include the users of the facility and the community at large, and has intensified the lack of clarity on the identity of the construction client (Newcombe, 2003).

Clients' involvement has long been recognised as an important factor in improving the industry's performance, and more specifically in stimulating innovation in design and construction (NEDO, 1975). Therefore, understanding what underpins and prompts clients' attitudes and actions is critically important for construction professionals in collectively taking the industry forward (Boyd and Chinyio, 2006). Identifying construction client types is a first step in achieving such understanding.

This chapter initially presents a discussion of the concept of taxonomy. Subsequently, the term 'client' is defined, and the different types and classifications of clients as proposed in the literature presented. A proposed taxonomy for the construction client is then presented, based on both the literature and on discussions undertaken by the CIB (International Council for Research and Innovation in Building and Construction) Task Group 58 workshop held in January 2007. Finally, questions are put forward for further research.

7.2. Taxonomy

The term taxonomy originally referred to the science of classifying living organisms. The word was later applied in a wider sense, emerging as a classification of things, or the principles underlying the classification (Murphy, 2002). Most types of physical or conceptual entities (products, processes, knowledge fields, human groups, places, etc) may be classified according to some taxonomic scheme at any level of granularity (Rasch, 1987).

Taxonomy is, therefore, a theoretical operation in which groups, classes, or sets are systematically organised and linked according to some criterion (Mayr, 1982; McKelvey, 1982). Taxonomies may be hierarchical or non-hierarchical, describe network structures, and can be derived inductively or deductively (Rasch, 1987).

Taxonomy development involves determining a classification scheme, i.e. identifying differences, attributes or properties on which to base a classification, and the techniques used to construct it (McCarthy, 1995). As it includes systematically organising concepts and criterion links, taxonomy may be considered as a conceptual framework.

The next section synthesises the definition of the term 'client' and client types as proposed in the literature.

7.3. The construction client

7.3.1. Definition of the term 'client'

Different approaches have been adopted to understand the identity of the construction client. The current poor understanding of who the construction client is begins from the lack of clarity in the definition of the term 'client' (Boyd and Chinyio, 2006).

The term client has connotations of an individual as a result of historical and other uses. Consequently, early views tended to assume that 'client' implies a person or a well-defined group of people, which act as a single entity (Bertelsen and Emmitt, 2005). For instance, clients have been defined as the person or firm responsible for commissioning and paying for the design and construction of a facility (BPF, 1983). Thus, clients were seen as the initiators of projects and those that contract with other parties for the supply of goods and construction services.

In construction it is, however, unlikely that a client will be an individual; most clients are organisations or groups of people (Bertelsen and Emmitt, 2005). Organisational factors, rather than an individual's decision, tend to influence the client's decision to commission a project (Kamara *et al.*, 2000). Even when the client is legally an individual, the client's family and associates are part of the decisions surrounding the development, as they have a stake in the project. Furthermore, a client can be a representative of the owner, and act with delegated authority on the owner's part (Boyd and Chinyio, 2006). Therefore, organisational influences create complexity in determining the identity of the clients and their requirements, to allow different perspectives to be considered (Kamara *et al.*, 2000).

Cherns and Bryant (1983) observed that project team members generally do not appropriately acknowledge the complexity of most client organisations. They also state that construction professionals sometimes seem impatient with this complexity, tending to erroneously assume that the client clearly knows what his/her requirements are.

Kamara *et al.* (2000) further stress the importance of understanding that a constructed facility is not an end in itself, but a means to satisfying the business needs of the client. This creates the need to understand different perspectives represented by the client as well as to translate business needs in construction terms.

Furthermore, Flanagan *et al.* (1998) points out that this issue is further complicated as different people tend to have different clients or customers throughout the supply chain within a project; for example, the development company sees the investors, the financers and the tenants as the client, while the specialist contractors and material component suppliers see the contractor as the client.

In summary, construction clients are often complex and multifaceted in nature, comprising several interest groups whose objectives differ and may be conflicting (Green, 1996; Boyd and Chinyio, 2006). The existence of multiple stakeholders with different agendas, viewpoints, needs and interests has introduced political factors into the design process (Green *et al.*, 2004). This influences the requirements management, resulting in the need to negotiate a shared understanding of clients' requirements as well as the clients' business needs. Therefore, it is difficult to define 'clients' within the complex context of individuals and organisations that have interests in the delivery and use of constructed assets. It is true, however, that broad types of clients have been identified, described in the next section.

7.3.2. Types of clients

Previous research has recognised types of clients, suggesting diverse ways in which they could be classified. Different terms have, however, been applied to refer to similar classifications, which contributes to the existing poor clarity on the clients' identity.

Darlington and Culley (2004), for example, described the 'identifiable' and 'virtual' customer. The identifiable customer is represented by the individual who has a specific design problem, for example, a family needing a new house. The individual has a clear view of the problem and of its context, which can be discussed directly with those responsible for developing design. By contrast, the 'virtual' customer represents a class of individuals who might be satisfied by some product, the design of which will satisfy a collection of requirements. The actual process is quite different in capturing and managing requirements for each type of client. Most design methodologies consider requirements' capture for the 'virtual' customer (e.g. Cross, 2000). Furthermore, British Property Federation (BPF, 1983) definition of the client may be appropriate for the 'identifiable' customer, but it belies the multifaceted nature and complexity of the 'virtual' customer.

Similarly, Zeisel (1984) has emphasised the difference between *'paying clients'* (an entity/organisation signing-off documents) and *'end users'* (as wider functional clients). 'Paying clients' can be approached as the 'identifiable' customers, while 'end users' could be seen as 'virtual' customers. Edmondson (1992) has also referred to these two types of clients as the *'apparent customer'* and *'the user'*.

It is important to emphasise that, usually, a relatively clear relationship and exchange of information exists between paying clients and designers. However, there can be a lack of understanding between clients and end users, as well as users and designers. This happens as, in some instances, it is difficult to clearly establish who the end users will be, as these may include individuals as well as corporate entities or communities. However, the views of both clients and users need to be considered for successful outcomes (Zeisel, 1984).

A second criterion identified in the literature to classify clients focuses on the level of client experience with construction. Higgin and Jessop (1965) have distinguished between *'sophisticated'* and *'naïve'* clients on the basis of their previous experience with construction. Boyd and Chinyio (2006) include *'partially informed'* clients in this classification, i.e. those who have procured a few projects.

Masterman and Gameson (1994) categorised clients as: 'primary and secondary', 'inexperienced and experienced', generating four categories. The authors drew from Nahapiet and Nahapiet (1985), in which 'primary' clients are seen to be those whose main income is derived from constructing buildings (i.e. property developers), and 'secondary' clients are those who build to perform other business activity (e.g. healthcare). Masterman and Gameson (1994) further argued that clients cannot be classified solely on the grounds that they possess previous experience; they must have experience of the particular building type in question.

Hillebrandt (1984) referred to 'continuing' and 'one-off' clients, while Cherns and Bryant (1983) made a distinction between 'unitary' and 'pluralistic' clients. Similarly, Blismas *et al.* (2004) emphasised the difference between clients 'involved with one-off projects' and those who have 'large ongoing construction portfolios'. The same authors further argue that the single project paradigm dominates the construction management literature; however, this does not accurately reflect the reality of many clients who are effectively involved in multi-project environments.

Construction clients have also been classified according to the sector in which they operate. Hillebrandt (1984) stressed differences between 'public' and 'private' sector clients. NEDO (1975) highlights that, in the 'public' sector, the expression of the requirements to meet social needs is seldom clear cut, therefore determining requirements and value is challenging. Furthermore, as the 'public' sector represents 40% of construction orders in the UK, it is believed that it could make a contribution towards the improvement of construction by demonstrating '*it is a best practice client*', which consistently secures the best whole life performance that the industry can offer (Strategic Forum, 2002). However, this approach has been contested, as some authors advocate that the client should not be seen as responsible for the improvement of the industry (Ivory, 2005).

Boyd and Chinyio (2006) extended this categorisation to include 'public', 'private' and 'mixed'. 'Public' was divided into national and local, whereby services are provided and controlled nationally or locally. 'Private' was subdivided into industrial and services. The final division, 'mixed', involves a degree of public and private enterprise, subdivided into not-for profit and private regulated organisations.

The literature also presents classifications based on the type of project the client is involved with. For instance, Boyd and Chinyio (2006) described a 'client system-based' and a 'client need-based' categorisation. The 'client system-based' emphasises the nature of organisation and the source of project finance, while the client need-based focuses on the built asset form, use and ownership type, being, therefore, a project/product classification approach.

Cherns and Bryant (1983) stressed the importance of considering the environment in which the client operates. For instance, if the client's business environment is stable, then requirements are likely to remain relatively constant over time; however, if this environment is dynamic, then requirements may well change as design evolves. Such consideration may help construction professionals to predict the level of changes and consequently consider appropriate approaches to manage client requirements as well as the level of, and need for, clients' involvement in the process.

Finally, the scale of the client organisation has also been used as a means to differentiate between types of clients, as larger organisations tend to procure larger amounts on a more regular basis (Boyd and Chinyio, 2006). The Strategic Forum (2002) describes that large, repeat clients normally have in-house teams and processes which ensure they establish requirements at the project outset. However, small and occasional clients do not have such structures in place and are perceived to need guidance on the practical steps to be taken when commissioning a project.

In conclusion, six generic criteria to classify clients were identified in the literature, as described in Table 7.1. These are as follows: (a) paying clients and users; (b) level of experience of the client with construction; (c) nature of the client organisation – sector in which it operates; (d) type of clients' business; (e) size of the client organisation; (f) rate of change in the clients' organisational environment.

It is important to note that most client type descriptions in Table 7.1 reflect extreme representations. For some of the types presented, however, there is, in reality, a continuum in which the client could be located, i.e. clients may be completely inexperienced in construction, or they may have some level of experience with the industry but maybe not with the project type, or they may be experienced with the specific project type. Such continuum can also be identified for the rate of change in the client organisation and for the size of the client organisation.

In summary, the literature presents different approaches to describe the construction client's identity. Different terms have been used to refer to similar classifications, which may have contributed to the poor clarity about the construction client. Furthermore, the classifications identified in the literature are too generic to effectively support a better

Table 7.1   Summary of client types as identified in the literature

Client 'type'	Terms used to refer to the type in the literature
Paying client and users	• Apparent customer and user[a] • Paying clients and end users[b] • Identifiable customer and virtual customer[c]
Level of experience with construction (and level of experience with specific building type)	• Sophisticated, partially informed and naïve[d,e] • On-going portfolios and one-off projects[f,g] • Unitary and pluralistic[f]
Nature of the organisation – sector	• Public (national and local)[h,i,l] • Private (industry and service)[h,i,l] • Mixed (mix of public and private enterprise; not-for profit and private regulated)[e]
Clients' business type	• Property developers (primary) and those who build to perform some business activity (secondary)[j,k]
Product type	• Building form types, building use types, ownership types[e]
Rate of change on environment	• Static versus dynamic environments[f]
Size of the client organisation	• Small, medium or large companies[e,l]

[a] Edmondson (1992). [b] Zeisel (1984). [c] Darlington and Culley (2004). [d] Higgin and Jessop (1965). [e] Boyd and Chinyio (2006). [f] Cherns and Bryant (1984). [g] Blismas *et al.* (2004). [h] Hillebrandt (1984). [i] NEDO (1975). [j] Masterman and Gameson (1994). [k] Nahapiet and Nahapiet (1985). [l] Strategic Forum (2002).

understanding of the client. It is, however, challenging to determine more detailed clients' characteristics, as the clients' base is diverse.

7.4. A proposed taxonomy for the construction clients

The literature has enabled the identification of *attributes or properties* on which to base the construction clients' taxonomy, as described on Table 7.1. However, some have questioned the relevance of a taxonomy to support innovation in construction.

Green (1996) has argued that understanding clients according to their type is naïve as client organisations are *not* unitary entities whose objectives are clear and consistent over time. The author further states that 'construction professionals ... must also accept that there will never be any single interpretation of a complex multifaceted client' (p. 157). Following such an argument, it is, therefore, necessary to understand the characteristics of the specific client organisation and the ways in which the client makes sense of his own organisation when developing a project. Clients' needs will, therefore, be identified via an evolving relationship between construction professionals and building clients/users. Therefore, a taxonomy of clients may not serve the

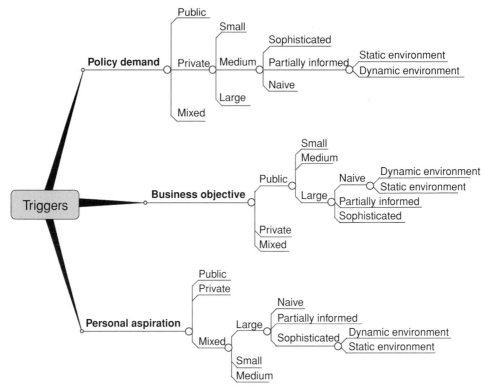

Figure 7.1 Construction clients taxonomy.

purpose of better identifying and responding to client needs, or in better understanding the clients' role in innovation.

Furthermore, Table 7.1 makes it clear that existing classifications have not sufficiently focused on the clients' needs. Even though these may clarify that different stakeholders' needs should be considered (e.g. paying clients and users), it does not support the identification of what these needs may be. Identifying clients' needs is a basis through which their motivations and willingness to innovate could be recognised.

We argue here that a typology can only be effective if it is based on a clients' needs-oriented model. This model should emphasise what triggers the client to become involved in construction, i.e. personal aspirations (e.g. a person building a house), business objectives (e.g. BAA building an airport terminal) or policy demands (e.g. a PCT developing a healthcare facility). Consequently, it would allow the identification of how a client organisation may interact in the decision-making process in a project, and hence how they may influence innovation.

Figure 7.1 presents the proposed construction clients' taxonomy. The taxonomy starts from what triggers the clients' involvement in construction. It then clusters different types of clients according to the sector in which the client organisation operates, its size, its level of experience with construction, and finally the level of change in the business environment in which it operates.

It is believed that the taxonomy should *effectively* support a better understanding of the construction clients and their needs, attitudes and roles in the construction process.

It is important to consider that the level of motivation of the client to support innovation in construction may be low or high for each of the proposed triggers.

7.5. Questions for further research for clients driving innovation

Based on the discussion presented above, questions for further research are posed, discussed in the following.

7.5.1. How to establish the client's motivation to innovate in design and construction?

Ivory (2005) reported on the potentially negative implications of the client upon design innovation. The author presents empirical evidence from three case studies in which the client had a desire to avoid the risk associated with innovation; for example, late or over budgeted projects, or higher than expected running costs or maintenance costs. Therefore, it is possible to state that in some cases clients may not see the benefits from innovation to themselves, so they would have no reason to support it.

Clients have a disproportionate share in the risk of innovation in buildings; therefore client-led innovative construction projects should be seen as exceptions rather than the norm (Ivory, 2005). It is then necessary to consider the following question.

7.5.2. What type of clients, in what circumstances, tend to support innovation?

Assuming the need to understand the role of the construction clients in driving innovation, one should first understand what motivates a client to support or impinge innovation in construction. Empirical research is needed to identify links between types of client and their willingness to innovate. In the literature there seems to be an underlying assumption that large complex public sector clients are more likely to support innovation than small private sector client; however, this assumption needs to be validated.

7.5.3. How to identify the clients' ability to be engaged in construction and support innovation?

There is a need to question research assumptions when dealing with construction clients. Past research identified that clients can have difficulties in providing timely and appropriate information for design (e.g. Darlington and Culley, 2004). Such difficulties are also recognised by the industry, and many professional reports are dedicated to provide guidance to clients (e.g. CABE, 2003), NHS Estates Best Client Guide (NHS Estates, 2002) and reports from the Construction Clients' Group and the Strategic Forum's Accelerating Change (Strategic Forum, 2002).

Nevertheless, academic recommendations appear to be somewhat prescriptive and assume a high level of client knowledge. Figure 7.2 presents a summary of activities clients should undertake to achieve successful building outcomes (see Tzortzopoulos

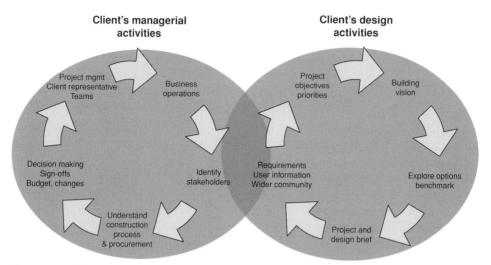

Figure 7.2   Clients' activities in the design and construction process.

et al. 2006 for details), including managerial and design activities. Managerial activities involve the (re)design of clients' business operations, as well as the definition of project management structures and decision-making processes. Design activities are those necessary to provide the design team with information to develop design solutions, i.e. the definition of the building vision, project priorities and appropriate building requirements.

Furthermore, there seems to be a clear assumption that the 'clients should enter the construction process with a clear understanding of their "business" needs and their environmental and social responsibilities and hence the functionality they require from the finished product' (Strategic Forum, 2002, p. 20). However, no other industrial sector seems to have such expectations from its clients. For instance, when buying a computer, the client does not expect or want to get involved in its manufacturing. In fact, it is usually a burden for the client organisation to get involved with the construction industry. This needs to be taken into account by the industry.

7.6. Closing remarks

Possible solutions for the questions put forward in this chapter may emerge from a better understanding of the epistemological basis through which we try to understand the construction client.

Most typologies in the literature assume a somehow positivistic approach in which the understanding of the clients seems to be based on an 'objective' and 'stable' client structure and composition. On the contrary, client organisations may be approached from a phenomenological perspective, in which they are seen as coalitions composed by different individuals with varied and sometimes conflicting objectives. Therefore, *the clients* may be seen as clear and objectively defined entities or they may be seen as groups that understand their world subjectively through action, and therefore interpret and act on the design and construction process subjectively. In such a view, their understanding of the (organisational) problem they deal with evolves, based on their actions and subjective interpretation of the information made available throughout design and construction.

One may argue that there is probably an 'objective' part of the problem definition, and a 'subjective' part which constantly evolves. Similarly, there may be objective definitions of client organisations characteristics/types that may support the industry to predict the likelihood of their involvement in the process and support for innovation. Therefore, it may be possible to identify causal links between client types and their interaction and influence over the project. The taxonomy here proposed is aimed at addressing such an objective definition. However, the taxonomy must be continually interpreted within specific contexts and situations.

References

Bertelsen, S. and Emmitt, S. (2005) The client as a complex System. *Proceedings of the 13th Annual Conference of the International Group for Lean Construction*, Sydney, Australia, 73–79.

Blismas, N.G., Sher W.D., Thorpe, A. and Baldwin, A.N. (2004) Factors influencing project delivery within construction clients' multi-project environments. *Engineering Construction and Architectural Management* **11** (2): 113–125.

Boyd, D. and Chinyio, E. (2006) *Understanding the Construction Client*. Oxford, UK, Blackwell Science.

British Property Federation (BPF) (1983) *Manual of the BPF System for Building Design and Construction*. London, British Property Federation.

Commission for Architecture and the Built Environment (CABE) (2003) *Creating Excellent Buildings: A Guide for Clients*. Report, CABE.

Cherns, A. and Bryant, D. (1983) Studying the client's role in construction management. *Construction Management and Economics* 2: 177–184.

Cherns, A.B. and Bryant, D.T. (1984) Studying the client's role in construction management. *Construction Management and Economics* 2 (2): 177–184.

Cross, N. (2000) *Engineering Design Methods: Strategies for Product Design*, 3rd edn. Chichester, John Wiley and Sons.

Darlington, M. and Culley, S. (2004) A model of factors influencing the design requirement. *Design Studies* **25**: 329–350.

Edmondson, H.E. (1992) Customer satisfaction. In: *Manufacturing Systems: Foundations of World-Class Practice* (eds J. Heim and W.D. Comptom). Washington, DC, National Academy Press.

Flanagan, R., Ingram, I. and Marsh, L. (1998) *A Bridge Top the Future: Profitable Construction for Tomorrow's Industry and its Customers*. London, Thomas Telford.

Green, S. (1996) A metaphorical analysis of client organisations and the briefing process. *Construction Management and Economics* **14**: 155–164.

Green, S., Newcombe, R., Fernie, S. and Weller, S. (2004) *Learning Across Business Sectors: Knowledge Sharing Between Aerospace and Construction*. Research Report, University of Reading.

Higgin, G. and Jessop, N. (1965) *Communications in the Building Industry: The Report of a Pilot Study*. London, Tavistock.

Hillebrandt, P.M. (1984) *Analysis of the British Construction Industry*. Basingstoke, Macmillan.

Ivory, C. (2005) The cult of customer responsiveness: is design innovation the price of a client-focused construction industry? *Construction Management and Economics* **23**: 861–870.

Kamara, J.M., Anumba, C.J. and Evbuomwan, N.F.O. (2000) Establishing and processing client requirements—a key aspect of concurrent engineering in construction. *Engineering Construction and Architectural Management* **7** (1): 15–28.

Masterman, J.W.E. and Gameson, R.N. (1994) Client characteristics and needs in relation to their selection of building procurement systems. *Proceedings of CIB W96 Symposium*, 12–13 November, Hong Kong, pp. 79–87.

Mayr, E. (1982) *The Growth of Biological Thought: Diversity Evolution and Inheritance*. Cambridge, MA, Harvard University Press.

McCarthy, I. (1995) Manufacturing classification: lessons from organizational systematic and biological taxonomy. *Integrated Manufacturing Systems* **6** (6): 37–48.

McKelvey, B. (1982) *Organizational Systematics: Taxonomy Evolution, Classification*. Berkeley, CA, University of California Press.

Murphy, G. (2002) *The Big Book of Concepts*. Cambridge, MA, MIT.

Nahapiet, J. and Nahapiet, H. (1985) *The Management of Construction Projects: Case Studies From the USA and UK*. Ascot, Chartered Institute of Building.

NEDO (1975) *The Public Client and the Construction Industries: Building and Civil Engineering EDCs*. Report, Joint Working Party Studying Public Sector Purchasing, Building Engineering Economic Department Committees, HMSO, London.

Newcombe, R. (2003) From client to project stakeholders: a stakeholder mapping approach. *Construction Management and Economics* **21** (8): 841–848.

NHS Estates (2002) *The Best Client Guide: Good Practice Briefing and Design Manual*. NHS Estates publication. Available at http://www.nhs-procure21.gov.uk/downloads/best_client_manual/best_client_clinical.pdf.

Rasch, R.F.R. (1987) The nature of taxonomy. *Journal of Nursing Scholarship* **19** (3): 147–149.

Strategic Forum. (2002) Accelerating change. *Strategic Forum for Construction*, London. Available at http://www.strategicforum.org.uk/pdf/report_sept02.pdf.

Tzortzopoulos, P., Cooper, R., Chan, P. and Kagioglou, M. (2006) Clients' activities at the design front-end. *Design Studies* **27** (6): 657–683.

Zeisel, J. (1984) *Inquiry by Design*. Cambridge, Cambridge University Press.

8 Clients' roles and contributions to innovations in the construction industry: when giants learn to dance

Charles Egbu

8.1. Introduction

Innovation, which is generally viewed as the successful exploitation of an idea, which is new to the unit of adoption, is a complex social process. Understanding the complexity of innovation is useful in understanding the contributions and the role that clients play (or could play) in innovation in the construction industry. The market place 60 years ago is different from the market place of today. Few would argue that the business environment in which clients and construction organisations face today is complex, chaotic and dynamic. Technological shifts, globalisation, the onset of a new economic system based on knowledge rather than capital, along with the creation of economic value by integrating corporate social and environmental responsibility issues now increasingly pose real profound strategic challenges for businesses. These complex challenges involve numerous processes carried out and influenced by many stakeholders to set the tone and guide corporate level decisions. This is the backdrop in which clients are increasingly expected to impact on innovations. It would appear that waiting and expecting the contractor and other members of the project team to come up with innovations would not now suffice. Clients would increasingly need to be proactive in innovation in order to meet their exact needs and expectations. Although there have been a few studies that have attempted to consider some of the roles which clients can play in contributing to innovation (Nam and Tatum, 1997; Winch, 1998; Gann, 2003; Rose and Manley, 2005), there is still a paucity of research frameworks that allows one to better understand the key roles that clients play, which takes account of the nature and types of client, and the different types and nature of innovation. In the same vein, very limited empirical studies exist in the literature of the key factors that suppress a client's role on innovation, and how these could be effectively addressed.

It is important to stress again that innovation is a complex phenomenon, but despite diverse perspectives, researchers and practitioners have agreed on the importance of innovation as a pre-requisite for competitive advantage. Innovations come from many different sources and exist in many different forms. There is a dichotomy between radical and incremental innovation (Damanpour, 1987). Innovation can be radical, in response to crisis or pressure from the external environment, but it can also be incremental where step-by-step changes are more common. Moreover, a common typology distinguishes product and process innovation. Product innovation is seen to focus on cost reduction by obtaining a greater volume of output for a given input. Process innovation, on the other hand, describes new knowledge, which allows the production of superior quality output from a given resource.

Different writers also have different ideas about the amount of stages there are in an innovation process. A three-stage model incorporates idea generation, adoption and implementation (Thompson, 1965; Shepard, 1967) while Zaltman *et al.* (1973) have developed a 12-stage process model. Rothwell (1992) has developed an historical model tracing the evolution of innovation models since the 1960s. His 'five generations of innovation' model explicates the transition from the simple, linear models of the 1960s to the complex and interactive models of innovation. In the fifth generation model, innovation is perceived as a multifaceted process, which requires intra and inter-firm integration, through extensive networking. Similarly, Wolfe (1994) noted that '. . . Innovation is often not simple or linear, but is, rather, a complex iterative process having many feedback and feed-forward cycles' (p. 411). In essence, innovation can be viewed as a process of inter-linking sequences from idea generation to idea exploitation that are not bound by definitional margins and are subject to change. Therefore, it is necessary to understand the complex mechanisms of this process and the context in which the innovation takes place.

Research on innovation has developed and taken on various shapes over the last 60 years. The level of analysis in innovation research is a useful preparatory consideration. The individualist perspective, which is grounded in social psychology, is predicated on the assumption that the individual is the source of innovation. They are the 'champion[s]' (Madique, 1980) or 'change agents' (Rogers, 1983) in an organisation.

In contrast to the individualist perspective, it is postulated that the structures and functions of an organisation are the pivotal determinants of innovation (Zaltman *et al.*, 1973; Pierce and Delbecq, 1977). This structuralist perspective is grounded in open systems theory and structural contingency theory; therefore, organisations are analysed as systems of interdependent parts, which cannot exist autonomously. It is assumed that the organisational characteristics, such as size, strategy, longevity and function play a central part in organisational innovations (Zaltman *et al.*, 1973; Pierce and Delbecq, 1977). The structuralist perspective has been criticised for drawing inert conclusions about the nature of innovation and perceiving the organisation as an objective entity that is driven predominantly by predictable forces (Slappendel, 1996).

Increasingly there have been recommendations to take a more multivariate approach to the study of innovation. Integration of both the individual and organisational levels of analysis to achieve a synthesis between action and structure is encouraged (van de Ven and Poole, 1988). Attempts to incorporate these diametrically opposed concepts have influenced developments in process theory. In essence, process perspectives recognise the unpredictable and dynamic nature of innovation. It is a complex process with cognitive, social and political dimensions that should be understood in particular organisational contexts (Swan *et al.*, 1999).

8.2. Clients in the construction industry

It is generally true that he who pays the piper calls the tune. The clients in the construction industry play a significant role in the construction industry. They can be seen as 'giants' who would increasingly need to dance to the tune of modern times, especially with regard to their role in innovation. But who are the clients of the construction industry? One of the earlier definitions of client is the person or firm responsible for commissioning and paying for the design and construction of a facility (British Property Federation, 1983). It is, however, the case that some construction clients are groups of people or organisations. In their recent book, Boyd and Chinyio (2006) inform us that a client can be a representative of the owner or act with delegated authority of the owner. This definitional challenge whether a client is an individual or not raises a host of issues such as the decision-making powers of the client, the level of influence the client possesses and the requirements of the clients. There is also the issue of the nature and types of client. One divide differentiates a public sector client from a private sector client. Another divide differentiates a 'paying' client from an end user. There is also a school of thought, which makes the differentiation between an identifiable client and a virtual client. Clients can also be seen as one-off clients or repeat business/continuing clients, or sophisticated or naïve clients. Put simply, it could be argued that the nature and type of client is likely to impact on the role and contribution a client makes on innovation.

8.3. Developing a framework for improved understanding of the roles of clients in innovation

Given the complexity of innovations, the different types and modes of innovations and the nature and different types and categories of construction clients, it would be unwise to suggest that all clients play similar roles in different types of innovations, at different stages of innovation. A framework (Figure 8.1) could help us to understand what particular role a particular type of client could play at different stages of different types of innovation; and, what factors are likely to impact upon such roles.

In attempting to do this, this chapter draws on earlier studies conducted on innovative construction organisations and the roles played by clients (Egbu *et al.*, 1998a; Egbu, 1999). It also draws on a thorough review of literature on types and nature of innovations (Wolfe, 1994), and Models of Innovation (Rothwell, 1992). The author argues that not all clients are able to play equal roles in radical innovations or incremental innovations. For example, if we take the view that radical innovations are likely to be more risky than incremental innovations, then clients who have capabilities (including dynamic capabilities) in risk management are likely to be better placed in their role in radical innovations than those who are not. Similarly, different clients are likely to perform differently in product, process, service, technological or market innovations. In the same vein, the client's expectations and demands may impact upon the extent of their role in innovation. The same can be said about the role of clients in emergent, adopted/adapted or imposed innovations.

The seven key roles which clients play and could play in innovations are (Figure 8.1):

(1)   As a source/provider of knowledge (including customer capital) for innovation.
(2)   Effective leadership.
(3)   As a change agent (positively influencing project/organisational/industry structures, strategy, culture and reward systems; preventing innovation lag time;

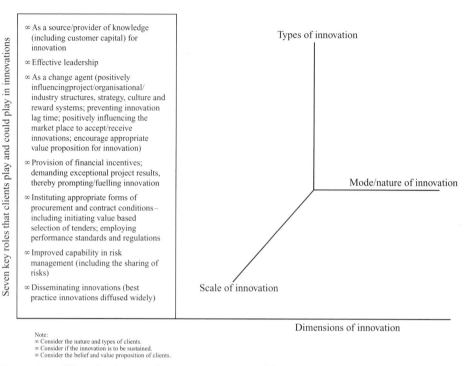

Figure 8.1   Clients' primary roles on innovation and key innovation variables.

positively influencing the market place to accept/receive innovations; encourage appropriate value proposition for innovation).

(4)   Provision of financial incentives; Demanding exceptional project results, thereby prompting/fuelling innovation.

(5)   Instituting appropriate forms of procurement and contract conditions – including initiating value-based selection of tenders; employing performance standards and regulations.

(6)   Improved capability in risk management (including the sharing of risks).

(7)   Disseminating innovations (best practice innovations diffused widely).

A thorough appreciation of the roles of construction clients in innovation would need to be considered within the context of types of innovation, scale of innovation and mode/nature of innovation (Figure 8.1). Figure 8.1 and Table 8.1 presented herewith should help us to shed more light on the different levels of contributions and impact which construction clients can make to innovations. It should also allow us to gauge where clients' impact is greatest, enabling us to investigate why that is the case.

8.4. The client and construction organisations in implementing innovations

Even where a client is taking the lead on innovation, effective implementation of the innovation within the project can be dependent upon the interplay between the client and the construction organisation.

Table 8.1   A framework for understanding the impact of clients' roles on key innovation variables

Primary roles which clients play in innovation	Impact of the primary roles which clients play on key innovation variables (very high, high, medium, low, very low)										Comments. To what extent does the nature and type of client positively impact on clients' role on innovation (very high, high, medium, low, very low)									
	Types of innovation					Scale of innovation		Mode/nature of innovation												
	Process	Product	Service	Market	Technology	Radical	Incremental	Emergent	Adopted/adapted	Imposed	Client as a person	Client as an organisation	Public client	Private client	Paying client	End user	One-off client	Repeat client	Sophisticated client	Naïve client
As a source/provider of knowledge (including customer capital) for innovation																				
Effective leadership																				
As a change agent – (positively influencing project/organisational/industry structures, strategy, culture and reward systems; preventing innovation lag time; positively influencing the market place to accept/receive innovations; encourage appropriate value proposition for innovation)																				
Provision of financial incentives; demanding exceptional project results, thereby prompting/fuelling innovation																				
Instituting appropriate forms of procurement and contract conditions – including initiating value-based selection of tenders; employing performance standards and regulations																				
Improved capability in risk management (including the sharing of risks)																				
Disseminating innovations (best practice innovations diffused widely)																				

The innovation process is dynamic. The dynamics of innovation, which have become increasingly intensive, result in high levels of risk and uncertainty arising, for example, from difficulties associated with accessing, transferring and assimilating knowledge, which is external to the organisation. These externalities include the heterogeneity of the knowledge sources that are important to innovation, and technological complementarities (including those between product and process innovations). Since innovations, especially radical or 'rule breaking' innovations, are associated with challenging thinking, unlearning as well as learning, entrepreneurial organisations appear to need general learning capacity. As Baden-Fuller (1995) and Senge (1990) have emphasised, the ability to learn from others, from the organisation around oneself and from one's own past, are critical elements in making progress.

In a study funded by the Economic and Social Sciences Research Council (ESRC), Egbu *et al.*, 1998a, 1998b, observed the following as some challenges which construction organisations face in exploiting clients'-driven innovations.

- Inability to link innovation strategy to the wider organisational business strategy.
- Managing the uncertainty and risks associated with innovation (e.g. risks associated with design and buildability of construction projects, technological risks, financial risks, contractual and increased exposure to litigious claims, safety risks, risk of complete failure of the innovation).
- Difficulties associated with being able to adequately scan and search the environments to pick up and process appropriate signals about potential innovation.
- Lack of resources and competencies associated with making strategic selections from potential alternatives and triggers for innovations and implementing the chosen innovation.
- Getting members of the organisation to 'buy-into' and support the innovation idea (commitment of the rank and file).
- Difficulty in getting the market (including subcontractors and suppliers) to take up the innovation (opportunism and the readiness of the market for the innovation).
- Difficulty in getting the Building Regulators to accept the innovation (e.g. design and the eventual product).
- Difficulty associated with auditing and measuring the benefits associated with innovation.
- Difficulty in understanding and putting in place an appropriate culture for innovation.

Construction organisations need to fully recognise the importance of knowledge creation, gathering, storing and exploitation. The culture and climate and the mechanisms in place should allow for the possibility for knowledge to be readily shared and transferred from project to project and across project teams. The important roles played by organisational culture and climate in construction innovations have been documented elsewhere (Egbu *et al.*, 1998a, 1998b). Construction organisations have a role and duty to respond to the needs of clients. They should also work in such a way as to encourage clients to exploit their innovation tendencies and capabilities. Exercising effective leadership is one way of doing this.

Leadership is an organisational responsibility. The value of institutional leadership is useful in the creation of structures, strategies and systems that facilitate innovation

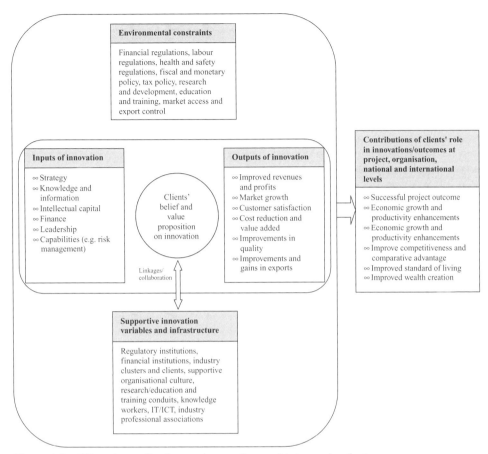

Figure 8.2   Clients' contributions to innovation and intervening factors.

and organisational learning. It should build commitment and excitement, collective energy and empowerment.

8.5. Contributions of clients to innovations

The role of construction clients in innovation can be significant, with contributions which can be felt at the project, industry, national and global levels. In this context, it is important to give cognisance to the varied factors that are likely to impact on transforming the clients' role in innovations in construction to the potential contributions this could have at project, industry, national levels and beyond. These factors are presented in Figure 8.2, taking a systems approach of input and output. Within all these, the value perception of the client is important. For example, there are clients who may see effective innovation as being derived from effective collaboration of key players as opposed to competition. There are those who may believe in openness in sharing appropriate knowledge for innovation as opposed to a closed approach. Similarly, there are clients who might see innovation from different lenses and perspectives (e.g. science, psychology, sociology, capital assets, resource-based perspective). Similarly, there are clients who may take a short, medium or long-term perspective of innovation.

8.6. Conclusions and recommendations

This chapter has documented the key roles that clients play in innovation, which includes leadership, the provision of an appropriate environment for innovation to prosper, financial incentive, risk minimisation. It argues that the contributions, which clients make to innovation can have project, organisational and national benefits. With innovations generally regarded as a means of introducing value to projects and businesses, clients not only have to dance to the tune of effective innovations but also increasingly need to be dictating the tune of the innovation music. It is noted that the nature and types of construction clients are likely to impact on their role and contributions to innovation. The client's value belief and proposition also play an important role on the extent of a client's role and contributions to innovation. A framework for better understanding the role of clients in innovation has been developed and documented.

References

Baden-Fuller, C. (1995) Strategic innovation, corporate entrepreneurship and matching outside-in to inside-out approaches to strategy research. *British Journal of Management* **6**: S3–S16. [Special issue, December.]

Boyd, D. and Chinyio, E. (2006) *Understanding the Construction Client*. Oxford, UK, Blackwell Publishing.

British Property Federation (BPF) (1983) *Manual of the British Property Federation System*. London, BPF. [Pub (01) 08.]

Damanpour, F. (1987) The adoption of technological, administrative, and ancillary innovation: impact of organizational factors. *Journal of Management* **13** (4): 675–688.

Egbu, C.O. (1999) Mechanisms for exploiting construction innovations to gain competitive advantage. *Proceedings of the 15th Annual Conference of the Association of Researchers in Construction Management (ARCOM)*, 15–17 September, Liverpool John Moores University, UK, Vol. 1, pp. 115–123.

Egbu, C.O., Henry, J., Quintas, P., Schumacher, T.R. and Young, B.A. (1998a) Managing organisational innovations in construction. *Proceedings of the Association of Researchers in Construction management (ARCOM) Conference*, 9–11 September, Reading, UK, Vol. 2, pp. 605–614.

Egbu, C.O., Henry, J., Quintas, P., Schumacher, T.R. and Young, B.A. (1998b) Cultural and managerial aspects of construction innovation. *Proceedings of the Third International Congress on Construction—Construction 21*, 21–22 May, Singapore, pp. 93–97.

Gann, D.M. (2003) Innovation in the built environment. *Guest Editorial* **21** (6): 553–555.

Madique, M.A. (1980) Entrepreneur champions and technological innovation. *Sloan Management Review* **21** (2): 59–76.

Nam, C.H. and Tatum, C.B. (1997) Leaders and champions for construction innovation. *Construction Management and Economics* **15**: 259–270.

Pierce, J.L. and Delbecq, A.L. (1977) Organization structure, individual attitudes and innovation. *Academy of Management Review* **January**: 27–33.

Rose, T. and Manley, K. (2005) A conceptual framework to investigate the optimisation of financial incentive mechanisms in construction projects. *CIB W92, International Symposium on Procurement Systems: The Impact of Cultural Differences and Systems on Construction Performance*, 7–10 February, Las Vegas, NV, Paper Ref. No. 125.

Rogers, E.M. (1983) Diffusion of innovations, 3rd edn. New York, Free Press.

Rothwell, R. (1992) Successful industrial innovation: critical success factors for the 1990s. *R&D Management* **22** (3): 221–239.

Senge, P. (1990) *The Fifth Discipline*. New York, Doubleday.

Shepard, H.A. (1967) Innovation-resisting and innovation-producing organizations. *Journal of Business* **40**: 470–477.

Slappendel, C. (1996) Perspectives on innovation in organizations. *Organization Studies* **17** (1): 107–129.

Swan, J., Newell, S., Scarbrough, H. and Hislop, D. (1999) Knowledge management and innovation: networks and networking. *Journal of Knowledge Management* **3** (4): 262–275.

Thompson, V.A. (1965) Bureaucracy and innovation. *Administrative Science Quarterly* **10**: 1–20.

van de Ven, A.H. and Poole, M.A. (1988) Paradoxical requirements for a theory of organizational change. In: *Paradox and Transformation: Towards a Theory of Change in Organization and Management* (eds R.E. Quinn and K.S. Cameron). Cambridge, MA, Ballinger, pp. 19–63.

Winch, G. (1998), Zephyrs of creative destruction: understanding the management of innovation in construction. *Building Research and Information* **26** (4): 268–279.

Wolfe, R.A. (1994) Organizational innovation: review, critique and suggested research directions. *Journal of Management Studies* **31** (3): 405–431.

Zaltman, G., Duncan, R. and Holbek, J. (1973) *Innovation and Organizations*. New York, Wiley.

9 Setting the game plan: the role of clients in construction innovation and diffusion

Kristian Widén, Brian Atkin and Leif Hommen

9.1. Introduction

Most research into innovation in the construction sector – and, for that matter, project-based sectors in general – has focused on the creation and development of innovations, with little emphasis on diffusion and dissemination (of those innovations) (Taylor and Levitt, 2005; Widén, 2006). The role of construction clients and their influence in the innovation process is another aspect that has received scant attention, although the general importance of clients – especially public clients and regular clients generally – has been stressed throughout the past decade (e.g. Egan, 2002). The role of the client regarding innovation has also been stressed in other industrial contexts. For instance, Porter (1998) argues that strong clients in the home market help companies to gain global, competitive advantage. In the construction sector, however, knowledge of how the client affects or might affect innovation and the diffusion of innovation is still an under-researched and theoretically weak area.

In construction, it is the client who chooses the process, procurement form and requirements, which to a large extent determines the boundaries for other actors in the sector. Certainly, it has been shown that the client can, by choosing appropriate procurement methods, provide incentives to other actors to be innovative and that, vice versa, innovation by other actors can prompt the client to change the procurement method (e.g. Briscoe *et al.*, 2004). Even so, the use of existing innovation theory to analyse the role of the construction client has received modest attention.

In this chapter, we argue that these gaps in knowledge stem from over-reliance on 'innovation process' models that have not been adapted to the peculiarities of the construction sector or, for that matter, project-based sectors in general. As a remedy, we recommend that 'systems of innovation' (SI) models be used to develop more adequate analyses of the client's role by taking into account relevant aspects of the sectoral and national context.

9.2. Current innovation framework

There is an increasing understanding among scholars that linear models of the innovation process do not capture the complete picture of innovation in the knowledge-based economy (Van de Ven *et al.*, 1999). The innovation process is an '. . . iterative, cumulative and cooperative phenomenon . . . ' (Freel, 2003, p. 751) often with extra-organisational contacts. The interactive character of innovation is reinforced by the trend among companies to focus more on their core competence and, consequently, to rely on cooperation with others where any competence is missing. Thus, innovation tends to involve cooperation with external sources (Freel, 2003), which is not captured in linear models of innovation.

One major drawback of generic process models of innovation is that they do not take into account features of the industrial and institutional context that can have an important bearing on how the innovation process actually unfolds. In terms of industrial contexts, one issue that has emerged is the need for innovation theories to address industrial sectors other than traditional manufacturing, for which most existing theories have been developed (Hobday, 1998; Taylor and Levitt, 2005). Further, it is not only the organisational, sectoral and local context that will have an impact on the innovation process. There is also evidence that country-specific characteristics will be influential too (Miozzo and Dewick, 2002).

Various types of SI models – for example, complex products and systems (CoPS) and national systems of innovation (NSI) – have been developed to capture the impact of contextual factors on innovation processes. In a project-based sector such as construction, combining the NSI and CoPS frameworks may be more useful than simply relying on conventional process models for the study of innovation.

The NSI concept has been defined as the set of institutions and actors that play a major role in influencing innovative performance nationally (Nelson and Rosenberg, 1993). SIs are, in some respects, mainly focused on the development of knowledge and less on diffusion and the use of technology (Geels, 2004). According to (Edquist, 2004, p. 182), however, they encompass all the determinants of innovation processes, i.e. 'all important economic, social, political, organisational, institutional, and other factors that influence the development, diffusion and use of innovations'. Generally, SIs consist of two main components: organisations and institutions. Organisations are 'actors' or 'players' and institutions are the 'rules of the game', i.e. laws, common habits and norms (Edquist, 2004).

Taking a national perspective, the NSI can have important implications for construction innovation, including the role played by clients. Country-specific institutional set-ups can, for example, affect the adoption of new approaches to public procurement. Public–private partnerships (PPPs) have been widely adopted in many EU (and other) countries in order to 'establish a long-term interactive partnership between the public and private sectors' in the development and operation of new infrastructure (Riess and Välilä, 2005, p. 11). In Sweden, though, the central government has shown little interest in PPPs, with the result that they are almost entirely confined to initiatives by local government. There are no Swedish national policies, regulatory frameworks or guidelines regarding PPPs (Sveriges Kommuner och Landsting, 2005, p. 35). One consequence is that Swedish government bodies do not have the same opportunities to allocate risks amongst public and private partners as do their counterparts in countries where PPPs are widespread. This circumstance affects the character of their interaction with construction firms.

The CoPS concept can be situated within the broader framework of sectoral systems of innovation (SSIs) (McKelvey, 2001; Malerba, 2004). There are some characteristics that distinguish CoPS from traditional manufacturing industries. The former are changing networks of organisations that involve many different suppliers. Products tend to be one-off projects or small batches (Hobday *et al.*, 2000). CoPS is not merely an innovation system model, but also a model of production systems that continuously generate innovations.

Interaction with users is a key issue for successful innovation (Dodgson *et al.*, 2005), with most innovation literature focusing on clients as users (von Hippel, 1988). It is important to survey the needs and demands of potential users on a regular basis in order to match products to the market (Tidd *et al.*, 2001). The highly customised products of CoPS mean that it is often necessary to have user involvement in the design of the products or systems. In CoPS, however, the client/customer is not necessarily the end user, and the focus is on how to fulfil the demands of the client (Davies and Hobday, 2005). Clients often need to be skilled in systems design and architectural capabilities, such that innovation is enabled through intimate client–producer links. With the development of new technologies – those that are largely computer-based – communication of innovations between clients and producers has become easier (Dodgson *et al.*, 2005). Even so, new demands on the innovative capabilities of clients can be expected to arise out of the possible separation of client and user roles, combined with the emergence of new users with qualitatively different requirements. As a consequence of this kind of systemic change, both producers and clients may have to reconfigure their knowledge bases, capabilities and networks or find new ways of organising 'search activities', which may also have to take new directions (McKelvey, 2001, p. 327).

9.2.1. Innovation diffusion

Much literature on technical change and innovation focuses on creation and development, but it is not until an innovation is used, widely spread and diffused that any real gain is achieved (Stoneman, 2001; Hall, 2005). Since diffusion is the process by which the innovation is communicated through certain channels over time, amongst the members of a social system, it can take a long time to make an impact (Rogers, 2003). A key characteristic is that diffusion occurs within a social system or network. An increase in interaction in a social network may therefore increase the rate of diffusion (Pittaway *et al.*, 2004). The diffusion is affected by the network's structure and by people; for example, opinion leaders and change agents. The most prominent characteristic of opinion leaders is that they are more exposed to external communication. They have higher socioeconomic status and are more innovative than their followers. Change agents are often professionals with university degrees in technical fields (Rogers, 2003). In some areas, professional institutions or associations act as independent change agents, so-called innovation brokers (Winch, 1998).

Professional and trade associations, i.e. third parties operating in networks, are important for the development of informal relationships. The key feature of informal relationships is the transfer of tacit knowledge promoting learning (Pittaway *et al.*, 2004). Tacit knowledge, spread through informal communication channels, also provides an explanation for geographically confined diffusion (Cantwell, 2005). With increasing complexity of the problems to be solved, project outcomes will depend to a greater extent on professionally oriented communication (Swan and Newell, 1995).

Innovation brokers have been seen to negotiate innovations through the regulatory network (Winch, 1998) and to create new standards (Miller *et al.*, 1995). If need be, innovation brokers can provide guidance and knowledge to innovators and consult with other organisations in the sector. However, there is wide variation across NSIs with respect to the actors that may perform this role. Although professional bodies can be important in some countries, such as the UK (Winch and Campagnac, 1995), such innovation brokers are much weaker and less capable of playing a key role in Sweden and certain other countries (Bröchner *et al.*, 2002).

As diffusion is about communicating something new – to an extent unknown – there is always a degree of uncertainty. Potential advantage of an innovation triggers the need to learn more about the innovation. Most potential adopters do not evaluate innovations on the basis of scientific studies. They rely on subjective evaluations given to them by individuals in the same setting or context as themselves (Rogers, 2003).

Diffusion usually takes off slowly with a few adopters and increases gradually. It then increases at a faster rate until most have adopted the innovation and then it slows down again (Stoneman, 2001). Plotting the cumulative adoption portrays an S-shaped diffusion curve (Goldsmith and Foxall, 2003). An important factor explaining the slow start of the diffusion curve is the low relative advantage when the innovation is first introduced (Hall, 2005). The type of innovation affects its diffusion; for example, systemic innovations diffuse significantly slower than incremental innovations. Organisational variety, span (the number of specialised teams involved) and innovation scope impact on innovation diffusion. When organisational variety is high and the span of a systemic innovation increases to impact two or more specialist firms, extra coordination and control are required. Ways to overcome this problem are partnering and the co-location of cross-disciplinary teams (Taylor and Levitt, 2005).

Many variables will influence an organisation or individual when deciding whether or not to adopt or implement an innovation. Moreover, these variables will change over the diffusion curve (DeCanio *et al.*, 2000). This means that factors, which may convince an early adopter might have nothing to do with the factors convincing the majority later along the diffusion curve. Two general factors that were found to be important in a major IT adoption study for both early and late adopters, although with different characteristics, were infrastructure compatibility and supply side activities (Waarts *et al.*, 2002).

These aspects of diffusion indicate that 'lead' clients may play a central role in the diffusion of construction innovations. However, this role can vary significantly across types of innovation and different stages of the diffusion process, and also according to the kind of client. One type of client that is especially likely to have a strong influence on the construction sector is a large public body that repeatedly orders new buildings. The following part of our discussion will focus on this type of client and, more specifically, on how client roles and capabilities may vary according to the innovation objectives of public procurement.

9.2.2. Public procurement and innovation

Public procurement has long been recognised as one of the most effective mechanisms for stimulating innovation (Rothwell and Zegfeld, 1982). Arguments for its use as an innovation policy instrument refer to its use of the market power and technological competence of large public sector organisations to reduce risks, provide incentives, articulate demand and assure markets for producers who would otherwise hesitate to develop innovative solutions and products (Geroski, 1990). Moreover, public procurement can also play a vital role in the diffusion of innovations, where the spread of a new product or system to new markets or market segments typically requires further development work in order to adapt it to these new contexts (Edquist and Hommen, 2000, pp. 20–23). The distinction between 'direct' and 'catalytic' public pro-curement (the latter being carried out on behalf of other eventual end users) is also highly relevant for the diffusion of innovations (ibid.). Furthermore, it may be es-pecially relevant for the construction sector, given the frequent separation between 'clients' and 'users'.

Building upon work by Edquist and Hommen (2000) in classifying the public pro-curement of innovations, Hommen and Rolfstam (2007) elaborate a taxonomy that refers to three modes of user–producer interaction (adding 'cooperative' procurement to the 'direct' and 'catalytic' varieties) and three stages in the evolution of markets and technologies ('early', 'middle' and 'late'). Within the resulting matrix, represented in Table 9.1, 'cooperative' and 'catalytic' modes of interaction and 'middle' and 'late' evo-lutionary stages are especially relevant for the diffusion of construction innovations. Among other things, this taxonomy underlines the high potential for innovation in the later stages of market and technology development. It also highlights the need for effective coordination in innovation policies (e.g. environmental sustainability) that may require 'differentiated strategies' for the simultaneous use of several modes of public procurement.

Table 9.1 indicates the main categories of Hommen and Rolfstam's (2007) taxonomy, and also provides illustrative examples of selected cells within the matrix. For 'early stage, direct' procurement, the hypothetical example given suggests that the client–user will require both considerable buying power and a very high level of technical competence. For 'middle-stage, cooperative' procurement, the example corresponds to an empirical case in which the project failed, due to the conflicting rationalities, values and goals of multiple clients–users, suggesting a need for effective communication and coordination among partner organisations (Rolfstam, 2007). For 'late-stage, catalytic' procurement, the example again corresponds to an empirical case, where public author-ities, determined to avoid the failings of earlier generations of public housing, played a key role in focussing the expression of existing private demand and ensuring that es-sential requirements were met (Hommen and Larsson, 2004). These examples indicate considerable variation in governance and knowledge requirements across different types of procurement. A monopsonistic public agency pursuing its own priorities can behave quite differently from one that instead leads a group of buyer organisations with related but perhaps only partly overlapping agendas. When public agencies try to articulate demand on behalf of other potential end users, e.g. consumers, they must focus strongly on design criteria based on learning about user requirements through vertically extended networks.

9.3. Setting the game plan

As noted earlier, there are some important differences between the construction sector and manufacturing (i.e. mass-production sectors). The construction sector is project-based and is thus characterised by temporary undertakings to produce unique prod-ucts, thereby differentiating its context from that of traditional manufacturing. To add to the contextual differences, construction projects are usually carried out by many

Table 9.1 A taxonomy of public procurement of innovation

Modes of Interaction	Stages in the evolution of technologies and markets		
	Early ('fluid phase'; extensive product innovation)	Middle ('transitory phase'; focus on process innovation)	Late ('specific phase', minimal innovation by major producers)
Direct (single public organisation as both client and user)	Procurement by a public utility of a 'world-first' power generation plant using Hydrogen Fuel Cell technology		
Cooperative (multiple public and private organisations as both clients and users)		Public–private partnership for a local power plant using renewable energy sources	
Catalytic (single public client, acting on behalf of many private end users)			New public housing developments with specific 'social' objectives (e.g. integration or sustainability)

Source: Adapted from Hommen and Rolfstam (2007).

organisations, typically in the face of 'hard' boundaries between the different phases of the project as, for example, between design and production. Existing theories of innovation need to be contextualised and carefully tested before being applied to the construction sector (Widén, 2006). In the construction sector, new ideas are seldom adopted by the company, as in mass-production industries, but rather into specific projects (Slaughter, 1998; Winch, 1998). In contrast to manufacturing, the products of the construction sector are large, complex, long lasting and created by a temporary project organisation (Slaughter, 1998). The innovations often affect more than one organisation in the process making it harder for a single organisation to adopt something new (Miozzo and Dewick, 2004). Since the organisational context of the projects is defined through the choice of procurement and contractual forms chosen by the client it is clear that clients have a profound role to play in providing an organisational context in favour of innovation and innovation diffusion.

Research has shown that part of the construction sector may be considered as possessing the characteristics of CoPS (Gann, 2000). Others have used the framework developed in 'The Innovation Journey' (Van de Ven *et al.*, 1999) as a nonlinear cycle of divergent and convergent behaviours repeating themselves over time (e.g. Clausen, 2002). The main issue surrounding CoPS is that, by definition, it covers large and complex products where innovations are developed for the purpose of individual projects. The economic volume is large enough to cover the cost of development and implementation of the innovations. This is not the case in 'normal-sized' construction projects. Furthermore, diffusion to other projects or organisations is not dealt with to any great extent in the CoPS framework. Similarly, the innovation journey model focuses on the creation and realisation of innovations and not on how they diffuse. Both of these approaches do show, however, that clients need to be knowledgeable and experienced in order to drive or contribute to innovation. This suggests that construction clients who have the opportunity to be active in innovation and innovation diffusion are also those who offer repeat business.

Although diffusion is extensively discussed in the literature much of it is treated as a non-integral part of the innovation process. There is clearly a gap in innovation theory, and particularly diffusion theory, for project-based sectors such as construction (Taylor and Levitt, 2005). Nevertheless, existing theory has identified a number of important conditions for the diffusion of innovations of which context ranks highly. Communication with other organisations that depend on the nature of the innovation during its development can also be beneficial for its diffusion (Widén, 2006). Choosing procurement and contractual forms that enable sufficient communication is a key issue for clients. Apart from that, not much is known about the effects that clients, in general, and in construction, in particular, have or might have on the diffusion of innovations. However, studies of the public procurement of innovations – especially, cases of 'cooperative' and 'catalytic' procurement occurring in the 'middle' and 'late' stages of technology development – may contribute to our knowledge and understanding of this topic.

As stressed by SI approaches, the diffusion of innovations will, in some cases, be affected by the need for adaptation of existing regulations, norms or standards. In such cases, it may be necessary to use innovation brokers to negotiate the changes, both vertically and laterally in the supply chain as well as through the institutional framework; for example, regulations and industry norms. Innovation brokers may have an important role to play in those cases where it is not possible to engage in communication with other organisations during the development of the innovation. For instance, brokers may be able to aid in assessing how these other organisations will be affected and how to modify (if necessary) the development of the innovation. They may also be able to broker the innovation to those other organisations (Widén, 2006).

9.4. Conclusions

The literature regards the construction domain as significantly different from other industrial sectors to the extent that many accepted innovation theories cannot be directly applied and replied upon. However, it is also clear from the discussion in this chapter that certain conceptual approaches to the study of innovation are relevant for the construction sector. In the case of Sweden – and arguably many other countries – context is not adequately addressed by established theory and underpinning literature that mostly has roots in different cultures – a point that indicates the need for more serious consideration and application of the NSI approach. Similarly, theories of user–producer interaction and the 'lead user' concept are highly relevant. Since clients 'set the game plan' for the rest of the sector, with their choice of procurement and contractual forms, their actions need to be assessed.

There are a few things that can be said of existing theory. Communication across organisational boundaries is very important in general and especially for diffusion of innovations. This needs to be enabled, supported and reflected in clients' choices of procurement and contractual forms. There are many types of clients in construction and not all have the chance to drive innovations and innovation diffusion, because they lack the knowledge and experience to do so. For those that have the ability, it is important that they 'stand up to the challenge' and take that responsibility. From a policy perspective, it may be easier to get large public clients to do it, as studies of the public procurement of innovations suggest, than to force private clients into doing it.

The main issue remains, however, that there is some knowledge about how construction clients can affect innovation, but less about how they can affect innovation diffusion. Further research is needed to shed light on these issues. Since previous research has shown that, for project-based sectors, the development and diffusion of innovations need to be integrated, these aspects would need to be studied comprehensively if they were to yield worthwhile results.

References

Briscoe, G. H., Dainty, A. R. J., Millett, S. J. and Neale, R. H. (2004) Client-led strategies for construction supply chain improvement. *Construction Management and Economics* **22**: 193–201.

Bröchner, J., Josephson, P.-E. and Kadefors, A. (2002) Swedish construction culture, quality management and collaborative practice. *Building Research and Information* **30** (6): 392–400.

Cantwell, J. (2005) Innovation and competitiveness. In: *The Oxford Handbook of Innovation* (eds J. Fagerberg, D.C. Mowery and R.R. Nelson). Oxford, Oxford University Press.

Clausen, L. (2002) *Innovationsprocessen i byggeriet—Fra idé til implementreing i praksis*, BYG DTU.

Davies, A. and Hobday, M. (2005) *The Business of Projects—Managing innovation in Complex Products and Systems*. Cambridge, Cambridge University Press.

Decanio, S.J., Dibble, C. and Amir-Atefi, K. (2000) The importance of organizational structure for adoption of innovations. *Management Science* **46**: 1285–1299.

Dodgson, M., Gann, D. and Salter, A. (2005) *Think, Play, do—Technology, Innovation, and Organization*. Oxford, Oxford University Press.

Edquist, C. (2004) Systems of innovation: perspectives and challenges. In: *The Oxford Handbook of Innovation* (eds J. Fagerberg, D.C. Mowery and R.R. Nelson). Oxford, Oxford University Press.

Edquist, C. and Hommen, L. (2000) Public technology procurement and innovation theory. In: *Public Technology Procurement and Innovation* (eds C. Edquist, L. Hommen, and L. Tsipouri). Dordrecht, Kluwer.

Egan, S.J. (2002) Accelerating change, rethinking construction. *Strategic Forum for Construction*, London.

Freel, M.S. (2003) Sectoral patterns of small firm innovation, networking and proximity. *Research Policy* **32**: 751–770.

Gann, D. (2000) *Building Innovation–Complex Constructs in a Changing World*. London, Thomas Telford.

Geels, F.W. (2004) From sectoral systems of innovation to socio-technical systems—insights about dynamics and change from sociology and institutional theory. *Research Policy* **33**: 897–920.

Geroski, P.A. (1990) Procurement policy as a tool of industrial policy. *International Review of Applied Economics* **42** (2): 182–198.

Goldsmith, R.E. and Foxall, G.R. (2003) The measurement of innovativeness. In: *The International Handbook on innovation* (ed L.V. Shavinina). Oxford, Elsevier Science.

Hall, B.H. (2005). Innovation and diffusion. In: *The Oxford Handbook of Innovation* (eds J. Fagerberg, D.C. Mowery, and R.R. Nelson). Oxford, Oxford University Press.

Hobday, M. (1998) Product complexity, innovation and industrial organisation. *Research Policy* **26**: 689–710.

Hobday, M., Rush, H. and Tidd, J. (2000) Innovation in complex products and system. *Research Policy* **29**: 793–804.

Hommen, L. and Larsson, M. (2004) Linköping—mjärdevi science park and lambohov— regional report included as an appendix. In: *Social Inclusion and the Division of Labour I: The Geography of Growth Poles and Ethnic Niches* (ed M. Gray). Report presented to the third review of the research network on *Regional Impacts of the Information Society on Employment and Integration (RISESI)*, March, 2004, Brussels, Belgium, IST Directorate of the European Commission.

Hommen, L. and Rolfstam, M. (2008) Public procurement and innovation—towards a taxonomy. *Journal of Public Procurement* **8** (2).

Malerba, F. (ed.). (2004). *Sectoral systems of innovation: concepts, issues and analyses of six major sectors in Europe*. Cambridge, Cambridge University Press.

Mckelvey, M. (2001) Changing boundaries of innovation systems: linking market demand and use. In: *The Business of Systems Integration* (eds A. Prencipe, A. Davies and M. Hobday). Oxford, Oxford University Press.

Miller, R., Hobday, M., Leroux-Demers, T. and Olleros, X. (1995) Innovation in complex systems industries: the case of flight simulation. *Industrial and Corporate Change*, **4**: 363–400.

Miozzo, M. and Dewick, P. (2002) Building competitive advantage: innovation and corporate governance in European construction. *Research Policy* **31**: 989–1008.

Miozzo, M. and Dewick, P. (2004) *Innovation in Construction*. Cheltenham, Edward Elgar.

Nelson, R R. and Rosenberg, N. (1993) Technical innovation and national systems. In: *National Innovation Systems: A Comparative Analysis* (ed R.R. Nelson). Oxford, Oxford University Press.

Pittaway, L., Robertson, M., Munir, K., Denyer, D. and Neely, A. (2004) Networking and innovation: a systematic review of the evidence. *International Journal of Management Reviews* **5** (6): 137–168.

Porter, M.E. (1998) *The Competitive Advantage of Nations*. New York, The Free Press.

Riess, A. and Välilä, T. (2005) Editors' introduction. *EIB Papers—* **10** (1): 11–16. [Special issue on innovative financing of infrastructure: the role of public-private partnerships.]

Rogers, E.M. (2003) *Diffusion of Innovations*. New York, Free Press.

Rolfstam, M. (2007) Organisations and institutions in the public procurement of innovations: the case of the energy centre in Bracknell, UK. Paper presented at *DRUID-DIME Academy Winter PhD Conference on Geography, Innovation and Industrial Dynamics*, , January, Aalborg, Denmark.

Rothwell, R. and Zegfeld W. (1982) *Industrial Innovation and Economic Policy*. London, Frances Pinter.

Slaughter, E.S. (1998) Models of construction innovation. *Journal of Construction Engineering and Management* **124**: 226–231.

Stoneman, P. (2001) *The Economics of Technological Diffusion*. Oxford, UK, Blackwell.

Sveriges Kommuner och Landsting. (2005) *Offentlig-Privat Partnerskap: Lägesbeskrivning*. Stockholm, Sveriges Kommuner och Landsting.

Swan, J.A. and Newell, S. (1995) The role of professional associations in technology diffusion. *Organization Studies* **16**: 847–875.

Taylor, J.E. and Levitt, R.E. (2005) Inter-organizational knowledge flow and innovation diffusion in project-based industries. *System Sciences, 2005. HICSS '05. Proceedings of the 38th Annual Hawaii International Conference*, p. 247.

Tidd, J., Bessant, J. and Pavitt, P. (2001), *Managing Innovation*. Chichester, John Wiley & Sons.

Waarts, E., Van Everdingen, Y.M. and Van Hillegersberg, J. (2002) The dynamics of factors affecting the adoption of innovations. *The Journal Product Innovation Management* **19**: 412–423.

Van de Ven, A., Polley, D. and Garud, R. (1999) *The Innovation Journey*. Oxford, Oxford University Press.

Von Hippel, E. (1988) *The Sources of Innovation*. Oxford, Oxford University Press.

Widén, K. (2006) *Innovation Diffusion in the Construction Sector*. Lund, Division of Construction Management, Lund University.

Winch, G. (1998) Zephyrs of creative destruction: understanding the management of innovation in construction. *Building Research and Information* **26**: 268–279.

Winch, G. and Campagnac, E. (1995) The organization of building projects: an Anglo/French comparison. *Construction Management and Economics* **13**: 3–14.

10 Clients as innovation drivers in large engineering projects

Roger Miller

10.1. Introduction

The entrepreneurial function in large industrial projects is shared among clients, strategy consultants and systems engineering firms. Even when government agencies are sponsors, the entrepreneurial function needs to be shared if projects are to reach high levels of performance.

For the purposes of this chapter, clients are operators of large networks, such as electric power, roads and oil and gas that need to invest in substantial mission-critical systems to meet forecasted demand. As owners, clients expect to influence demand, as well as design choices through internal staff or consultants. However, clients generally lack the forefront knowledge required to plan alone; they thus engage in the generative search for innovative ideas with co-specialised players.

The concepts developed in this chapter are based on 10 years of research on large projects to identify factors affecting performance. A study of 60 large engineering projects around the world was followed by an analysis of 15 public infrastructure investments, with a focus on front-end decision making (Miller and Lessard, 2007). Then, 40 IT infrastructure projects were investigated in the banking and engineering sectors to gain an understanding of relations between clients and consultants.

This chapter is divided into five sections. First, large engineering projects are presented as games of innovation involving generative interactions between clients and co-specialised consultants. Second, the processes by which clients and consultants develop projects are described. Third, the skills needed by clients to sponsor and co-develop projects with consultants are examined. The fourth section describes shaping as an evolutionary process of progressively testing hypotheses over many episodes. A conclusion follows.

10.2. Large engineering projects as games of innovation

Large engineering projects are usually one-off ventures that introduce varying degrees of innovation into the productive system of the economy. In order to understand the role of clients in innovation in large engineering projects, we need first to compare them with other games of innovation.

Table 10.1 displays six archetypes of games of innovation (Miller and Olleros, 2007). The market-creation games are patent-driven discovery, system integration and platform orchestration. The market-development games are cost-based competition, large engineering projects and customised mass production. The horizontal axis of Table 10.1

Table 10.1   Archetypes of games of innovation

	Architecture		
Markets	**Self-contained modules**	**Tightly integrated systems**	**Open modular systems**
Market-creation processes	Patent-driven discovery	System integration	Platform orchestration
Turning points Market-evolution processes	Cost-based competition	Large engineering projects	Customised mass production

presents three types of value-creation and -capture exchanges according to three architectures of products and services. Buyers easily assess self-contained products, such as pharmaceuticals, medical devices and batteries, and select them on the basis of merits. By contrast, the game of Systems Integration involves extensive interactions over time between clients and producers. Finally, in modular open systems, value creation and capture is a co-evolutionary process of interactions between individual buyers, orchestrators and their eco-systems of complementary components suppliers.

The vertical axis is divided into two types of market processes: initial market take-offs, involving the combination of effervescent technologies and knowledge applications to new markets; and, evolving markets, characterised by highly structured demand with slow rates of growth and strong competition. Below, I give a short description of the six games to highlight the specificities of large engineering projects.

10.2.1. Patent-driven discovery

This is the world of relentless innovation. Markets pre-exist but new discoveries are needed to regularly replace products that lose their intellectual property protection. Large research projects in promising sectors enable the appropriation of returns in case of success. Industries involved in this mode of innovation include pharmaceutical and biotechnological products.

10.2.2. Cost-based competition

Self-contained products that have become commodities are built in large-scale facilities and distributed through broad distribution networks. Such systems offer opportunities for optimisation by blending IT, operational research and accumulated operating knowledge. Innovation amounts to solving concrete problems, replacing old assets with state-of-the-art facilities, and fighting substitution by developing new applications.

10.2.3. System integration

It involves the development of complex products or design tools based on advances in science and engineering. Clients are sophisticated users facing very stringent market requirements: they need tools to design airplanes, semiconductors or cars. Competitors in this game are specialised entrepreneurial firms. Close to 45% of sales are allocated to innovation. Versioning of products is due primarily to users' demand for continuously higher performance.

10.2.4. Platform orcshestration

Innovation in this game consists of orchestrating market creation with modular products such as personal computers or telecommunications products. Orchestrators operate at many distinct levels: the architecture of the platform and its principles; the complementary players offering independent innovation; interactions with buyers by launching prototypes to learn, test and improve products or services; and, promotion of collective initiatives and standards to foster interconnectivity and market development.

10.2.5. Customised mass production

Innovation in this game is aimed at avoiding commoditisation. Products such as cars or watches are individually tailored within a dominant design and assembled from mass-produced components. Mass markets are segmented according to the multiple dimensions of user needs and served with differentiated products or brands (style, touch and product features). Players are usually large (global) firms that grow networks of suppliers and have a worldwide marketing expertise.

10.2.6. Large engineering projects

Such projects focus on the design and implementation of mission-critical systems in sectors such as manufacturing, banking, financial services, power generation and communications. Clients are usually operators that have the expertise to evaluate, understand and even improve on choices. Projects end up as tightly integrated closed systems intended to perform complex tasks such as building large computers, nuclear plants or IT systems. Innovations emerge from interactions between operators and co-specialised consultants that understand the evolution of rapidly evolving infrastructure, such as information, communications and production technologies. Large engineering projects are usually undertaken in mature markets and focus on the building of closed, integrated product architectures.

10.3. Large engineering projects as joint innovations between clients and consultants

Large operators invest to face competitive challenges or opportunities but often lack the expertise to design entirely new capital-intensive systems. Clients want advice about market and technical uncertainties and about better ways to manage projects before they commit to substantial irreversible investments. For example, in the oil and gas field, clients of large projects know how to operate petrochemical plants but do not have the knowledge to design new state-of-the-art facilities. Similarly, telecommunications operators and banks rely on strategic advisors, system integrators and equipment builders to design and deploy wireless systems, financial transaction systems or Internet networks.

Innovations take place as the result of interactions between co-specialised experts. Typically, the process starts with interactions, at the senior management level, between large operators and strategy consultants to help to define competitive issues, opportunities and risks. The goals pursued are usually to radically transform cost structures, improve the delivery of new products, improve transactions with customers, or understand emerging market dynamics. Operators want to significantly improve productivity and coordination of material and information flows around the world.

Once the strategic issues have been decided upon, clients build a project team internally or with consultants to sketch out the architecture of the new system and outline specifications. They then invite systems engineers with accumulated expertise to imagine, design and articulate innovative solutions. In turn, specialised suppliers involved in implementation may do their creative work internally or engage in further generative interactions with their own contractors. New systems are thus integrated and assembled from closed, open or interoperable technologies and subsystems. Quite often, extensive outsourcing of activities is included.

The dominant logic of innovation is thus a shared process of problem definition, solution design and implementation. Demand influences the supply of solutions, but, in turn, solutions can reshape demand. The reputation of strategic advisers and the experience of systems engineering or IT firms are core selection criteria. Clients generally choose from a limited range of consultants and engineering suppliers nationwide or worldwide. In each sector, the same few names keep appearing on everyone's list of preferred partners. Novices find the going rough.

Innovations emerge from the sharing and shaping of ideas in generative debates. Consultants and systems engineers must stay significantly ahead of their clients to provide valuable advice. For instance, IT or engineering consultants invest around 15% of sales in R&D and capabilities building; 18% of staff time is allocated to innovation (Miller and Olleros, 2007). The means used by consultants to foster innovative competencies include working relationships with leading clients facing significant challenges; alliances with technology suppliers to understand expected new technologies; formalisation and codification of strategic or engineering methodologies; and, accumulation of past experience in archives, knowledge management systems or expertise directories. Many strategy or engineering consultants fund research institutes to explore and build scenarios about the future evolution of the sectors that they serve.

Clients, especially large operators, capture value through the improved effectiveness and efficiency that stem from ramped-up projects. Consultants capture value by gaining reputation, experience and new knowledge. System integrators capture value by building partnerships with top-level strategy consultants and clients, involving both of them in learning about platform evolutions. Bankers or investors involved in the financing of projects capture value by better risk management.

10.4. Competencies of clients

Clients are not equal in their competencies to engage in generative interactions with consultants, systems engineers and external parties. Some simply adapt to exogenously determined demand, while others see the interactions between demand and project solutions. Clients participate actively in all activities, from front-end concept shaping to implementation. Here are the bundles of knowledge, competencies and leadership abilities identified through research on projects.

10.4.1. Ownership competencies

The prime competency is the ability to build business models and forecasts on which projects can be based. Operators are in a position to develop demand forecasts with varying degrees of accuracy, estimate capital and operating costs, and make provisions for embracing the inevitable risks of projects. Clients behave as responsible owners that will not only operate the planned project for many years but also live with the consequences of choices. They quickly sort opportunities into categories and evaluate them based on prior experience and knowledge. They form task forces in which operations executives join with lawyers, consultants and investment bankers to decide rapidly whether project ideas have option value.

The shaping of projects is a costly business with high thresholds: 3–5% of the total cost for standard non-innovative projects, and up to 35% for complex opportunities that involve some degrees of innovation (Miller and Lessard 2001). Many executives or politicians view such expenses as frivolous because they involve negotiation, legal advice and community involvement. In contrast, effective clients have learned that such expenses are necessary to master risks.

10.4.2. Competencies for organising generative governance systems

Clients know that they do not have all the knowledge to develop projects alone; they organise governance systems. Their role is to force trade-offs and challenge knowledgeable consultants, system integrators and bankers. Clients with operating knowledge may achieve substantial innovations by interacting with consultants and suppliers that have appropriate levels of experience and knowledge.

Sponsors do not sit idle, waiting for probabilities to materialise; they judge risks, imagine ways to cope, and work hard to shape outcomes. Strategising about risks may start analytically, but it requires managers to quickly become experts in organisational science, diplomacy, law and public affairs. Strong uncertainties and indeterminacy are thus reduced by the use of repertoires of responses based on prior experience.

10.4.3. Competencies to absorb changes or kill projects

Projects often need to be reconfigured because of difficulties, technical opportunities or major events. Crises occur that require major changes. Superior solutions may appear and major risks may emerge. Owners need to be able to assess rapidly the benefits and costs of embedding unplanned changes. There is no point in sticking to planned solutions if high potential value will be overlooked. Should clients change priorities in the course of the shaping or implementation of a project, they may generate uncontrollable dynamics that could destroy the benefits of adopting changes.

Effective sponsors may also abandon projects when expected returns are not clear. In contrast, instead of rejecting unworthy project ideas rapidly, ineffective sponsors believe that projects can be reworked; they often find themselves having to kill projects after they and their partners have spent large sums.

10.4.4. Relational competencies

These are the skills involved in engaging in credible discussions with political authorities, bankers, multilateral agencies and other relevant parties. Clients need to be reputable enough to enlist pension funds, development banks or local partners, and to forge long-term alliances that will be able to watch over and support projects and save them from risks.

10.4.5. Resources for survival when downside risks materialise

Sponsorship coalitions are particularly important at the ramp-up stage, when restructuring may be necessary. Successful clients have staying power based on the operation of assets or revenue streams from a diversified portfolio of projects. Revenue streams make it possible to fund adequate development of front-end choices, react flexibly in times of crisis and call on a network of partners and coalition members to help a project survive.

10.5. Approaches to the shaping of projects

Large engineering projects are certainly difficult technical tasks, but they are primarily complex managerial and socio-political challenges. Achieving problem solving and coordination of the interests and contributions of each powerful party is made possible by the design of governance systems. The ability of sponsors to coordinate both demand and supply can lead to high value creation. Because of its dynamic and unpredictable nature, complexity has to be met by versatile approaches. Three types of management approaches may be used to achieve effectiveness in coordination.

10.5.1. Hyper-rationality approaches

Project-management theories view projects as ventures that can be planned and specified in advance (Cleland and Ireland, 2006). Hyper-rational approaches certainly recognise that concepts are open in early phases of projects. However, the need for closure makes planning necessary. A business case is built and experts are hired to design a solution, define specifications for work packages and select contractors using bidding processes. It is assumed that complications can be solved by engineering calculus, computations and better coordination. Uncertainty is manageable through investments in information; analytical tools can scope, reduce, and eventually minimise risks. Project changes are eventually fought in the legal system.

In this perspective, projects unfold over time but form a sequential unity. Each stage is a detailed definition of the previous one. The different stages are business case, preliminary design, detailed design, bidding and construction. Accelerating a project through construction management will increase costs to the point that there is a danger of sinking it. Proceeding with prudence will increase the risk of missing the opportunity that the project aims for.

While rational approaches are still valid for simple projects and construction phases, they are inadequate for complex projects facing dynamic futures. Value creation is lost partly because sponsors cannot define all issues in advance. Hyper-rational approaches assume that the future is probabilistic and that planning must be done early.

10.5.2. Adaptability and improvisation

A second approach stresses the need for adaptability and flexibility in the face of changes (Brown and Eisenhardt, 1998). The messiness of projects induces many observers to argue that they are basically unmanageable. Projects are described as ventures in which chance, political actions or improvisations dominate (Flyvbjerg *et al.*, 2005).

Expenditures are allocated to soft issues such as public affairs, but contentious decisions are avoided so as not to raise thorny issues and threaten the viability of projects. Politics and power are often more important than techno-economic analysis. Instead of collaborating to understand and solve problems, many players appear to be governed by opportunism. Decisions are never final but are remade, recast and reshaped. Oversights are common.

Without strong commitments, not much progress can be achieved. Flexibility is not always possible in large projects because of their indivisibility and irreversibility. Projects require choices to be made on thorny issues, thereby giving away many degrees of freedom. Commitment, not flexibility, is required to communicate credibility to affected parties or financial sponsors.

10.5.3. The perspective of evolutionary project shaping

Rather than evaluating projects at the outset based on projections of the full sets of benefits and costs over their lifetime, many sponsors view them in evolutionary perspectives (Miller and Lessard, 2007). Sponsors act as champions, actively shaping projects in response to emerging, changing and unexpected conditions. Projects are not planned in linear rational forms but are shaped in evolutionary manners to achieve high value (Metcalf, 2005). Projects involve multiple actors from many levels that need to coordinate their actions over a variety of episodes. Value is co-produced by means of dynamic, iterative and generative relationships.

Sponsors start with initial concepts that have the possibility of becoming viable. They then embark on shaping efforts and debates to refine, reconfigure and eventually agree to concepts that will yield value while countering risks. The seeds of success or failure are thus planted early and nurtured as choices are made. Clients cut their losses quickly when they recognise that a concept has little possibility of becoming viable.

Planning has been portrayed as ineffectual (Mintzberg 1994), but, paradoxically, clients are spending increasing amounts of resources and time on this activity as projects are shaped (Miller and Lessard, 2007). Projects are born as intuitive hypotheses that need to be confirmed. As issues arise and call for decisions, high levels of planning expenditures are necessary. Finally, concepts are formalised into viable configurations. Planning activities take place in the course of shaping activities.

A variety of intertwined issues have to be resolved one by one by clients, alone or in cooperation with partners or co-specialised firms. Should institutions be inappropriate, a large part of the shaping problem is to create substitute coalitions and eventually help to create new institutionalisable patterns. This typically involves 'buying in' some stakeholders and 'buying off' others. In some cases, the roles of stakeholders can be specified in advance. In many cases, though, it is not clear how to accommodate various interests, so the leading client must exploit the front-end period to identify a mutual-gains trajectory.

The shaping process combines deliberate actions with responses to emergent situations as projects progress through time. Managers introduce real-time mitigation strategies to influence chaotic situations. This, rather than selecting, is the essence of shaping.

10.6. The evolutionary shaping and steering of projects

Projects are innovations in themselves, as they introduce new capacities in productive systems and have both positive and negative effects. They involve multiple actors under the leadership of major sponsors. The degrees of innovation that they introduce range from marginal improvements to highly disruptive innovations. The more innovative projects usually call for extensive shaping activities.

The shaping of projects takes place over many episodes, which are not a sequence of stages but progressive issue resolutions. The path taken by each episode depends on previous ones but is to some degree autonomous and indeterminate. For each episode, clients form hypotheses about the extent of progress that can be achieved on the issues that need to be resolved. The costs of shaping are so high that sponsors first identify projects that stretch their capabilities but that, because of their complexity and risk, offer substantial benefits that could not be achieved with simpler, less risky undertakings. Planning and shaping efforts are expended to make projects economically viable, technically functional and socially acceptable.

Clients that build many projects over the years invest in the building of the capabilities of their partners and consultants. Inspired by the capability and maturity model in software engineering, they will make sure that co-specialised players have the capabilities to shape concepts through generative interactions (see Chapter 22). Let us look first at the structure of shaping episodes and then at their dynamic interrelations.

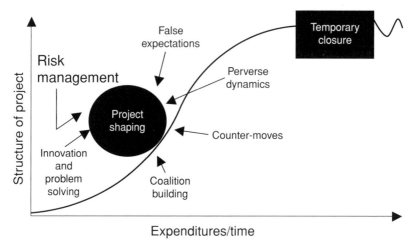

Figure 10.1   Shaping efforts to build momentum. (Source: *Strategic Management of Large Engineering Projects*. MIT Press.)

10.6.1. The structure of shaping episodes

Episodes start with momentum building, continue with reactions to opposing forces and end with closure. Each episode opens new options, closes old ones, and ends up with a particular concept configuration. At closure, clients and partners commit and thus lose some degrees of freedom. Figure 10.1 pictures shaping efforts as going up a hill through coalition building, problem solving and risk management.

10.6.1.1. Momentum building

Momentum is built by imagining concepts, promoting legitimacy and designing a configuration such that partners, affected parties and governments accept what is proposed. Unless sponsors have access to world-class cost estimates or benchmarks, they may err.

10.6.1.2. Meeting countering forces

The forces of criticism, calls for realism and professional challenges constrain shaping efforts and may even plant seeds of later failure or success. Clients that sponsor projects often yield to the temptations of unreasonable commitments by demanding that project solutions be developed fast. Excessive realism, in contrast, leads to scepticism and to the eventual rejection of good opportunities.

10.6.1.3. Conceptual closure

Each shaping episode ends with suggestions to abandon the whole project or to come to a temporary agreement on a conceptual configuration. Conceptual closure takes many forms: memorandum of understanding, business case, negotiated agreement, formal public commitment, sets of formal contracts and so on. Closure may be too early, too late, too rigid or too flexible. Key dimensions may have been omitted.

10.6.2. Evolutionary and procedural rationality

There are few projects in which major choices are made in a single episode. Investments are better viewed as staged options. Under the influence of pressure or legally empowered groups, project concepts may have to be redefined. Managers counter strong uncertainties and grapple with non-linear causalities, path dependence and feedback processes though progressive hypothesis testing. Successive shaping episodes eventually bring a project to full momentum and closure on an agreed-upon conceptual configuration. Herbert (2000) referred to this as procedural rationality. Substantive rationality focuses on selecting the optimal option, while procedural rationality recognises the unfolding of time, the presence of strong uncertainty and the need for collaboration.

Clients grope their way through an evolutionary decision-making process. Because of changing conditions that offer positive or negative consequences, each episode may start with a reconfiguration of the conceptual agreement achieved in the previous episode and pursue different paths. New hypotheses, options and challenges may lead to a reconfiguration of the problem. Clients need to have the ability to sense changes, estimate the benefits and costs of accepting changes and reconfigure resources to execute swiftly as shaping or execution activities go on.

Figure 10.2 illustrates the actual shaping efforts for a bridge project; five episodes characterise the progressive shaping from the initial hypothesis to construction. Many

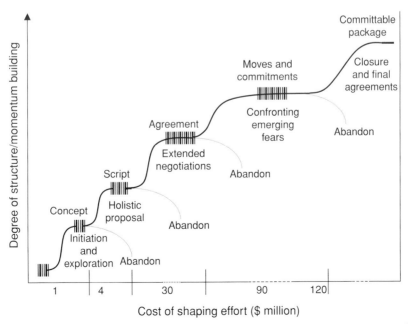

Figure 10.2   Crossing hurdles over many episodes. (Source: *Strategic Management of Large Engineering Projects*. MIT Press.)

projects do not go through all of these episodes because lock-in occurs early or sponsors kill them.

Early on in the project depicted in Figure 10.2, the client outlined a system of governance, a division of responsibilities and the necessary commitments. A fully developed script addressing pertinent risks and providing concrete solutions was built. The main shaping episodes were as follows.

10.6.2.1. Initiation and exploration

The initiation episode lasted one year (the range is usually 6–18 months) and closed when a credible party stated openly that he was ready to allocate funds and started debates on the ways and means of shaping and financing the idea. The infancy trap was avoided because the client honoured its commitments and agreed to coordinate its efforts with those of relevant specialised organisations.

Resources of a few million dollars were then assigned for exploration of the initial project hypothesis. Engineering, financial and marketing feasibility studies were undertaken. Conceptual closure was achieved after 18 months, when independent studies confirmed the viability of the concept. The output was a series of documents sketching out the initial concept but with an emphasis on technical issues. A position chapter was presented to legitimate authorities.

10.6.2.2. Development of a holistic proposal

Shaping interventions have the highest leverage during the early front-end period. The client/sponsor talked to regulators, public-affairs specialists, lawyers or bankers to arrive at a better understanding of issues. Progress was made on each issue by the following:

- Holding discussions and negotiations within a coalition of partners to create strong ownership.
- Reaching a satisfactory balance between ownership rights to build commitment to bear risks.
- Ensuring that investments are protected against opportunistic behaviours by focusing on devices such as long-term contracts and assumption of market risks.
- Gaining legitimacy through consent from affected parties by recognising rights, negotiation of compensation, and proactive strategies with communities.
- Achieving shock-absorption capabilities to handle crises by developing agreements, obligations, contingency funds and other means.

The holistic proposal was concerned with the building of an investment scenario and a business case for investors. Preparing the proposal was expensive. Specialised partners that had joined with the intent of contributing to overall value creation were also hoping to capture 'more than their fair share'. The result was a seesaw between positive-sum collaboration and zero-sum gaming. Eventually, the balance became acceptable.

10.6.2.3. Negotiations and issue resolution

Many issues cannot be resolved by promises or agreements. To confront emergent fears, clients often have to clean up polluted sites, organise referenda or build complementary facilities. They may also need to organise economic-development initiatives to demonstrate their credible commitments. Furthermore, numerous issues skipped in the holistic proposal require solutions through negotiation and problem resolution.

10.6.2.4. Confronting emerging fears

Facing social and environmental fears is a very expensive affair. Concrete moves to meet expectations and solve social and environmental issues have to be made. If parties are unable to forge agreements, they must wait for court or government decisions. The presence of public social- and environmental-assessment frameworks is extremely important in helping to solve dilemmas. Delays are the inevitable consequence of such formal assessments, but the public framework builds legitimacy and forces parties either to make trade-offs or to kill the project.

10.6.2.5. Closure on a committable package

Commitment on a final package takes place when all major issues have been resolved. In many projects, clients may have spent several hundred million dollars to gain consent, solve social and environmental issues and build agreements.

The construction sprint period begins once full shaping has been achieved. The value of speed is that revenues will arrive faster. Turbulence at this stage is usually lower, as major issues have been resolved. However, should clients change their minds and priorities, they will trigger dynamic processes that will be very costly.

The ramp-up period tests expectations about markets, technical functionality and social acceptability. In the case illustrated in Figure 10.2, after two years of ramp-up, the project performed satisfactorily.

10.7. Conclusion

Projects emerge in successive shaping episodes that start with the client forming a hypothesis about the progress that can be achieved on the issues that need to be resolved and the efforts that are required to develop strategies to bring resolution to issues. Nested issues need to be addressed first as well as the resources necessary to achieve progress towards closure.

Each episode opens new options and closes old ones until sponsors and partners achieve final lock-in, thus binding their commitments and losing most of their degrees of freedom. Shaping episodes start with momentum building, continue with countering opposing forces and end with closure. Shaping efforts push up hill through coalition building, problem solving and risk management in the face of counter-dynamics such as cynicism, false expectations and feedback effects.

Projects are made economically viable, technically functional and socially acceptable by the client and his consultants as a result of progressing on solutions to deal with the following issues: (1) negotiating a project proposition that truly creates value and can be progressively refined; (2) developing a stable business model for the project, to ensure that investments will be repaid and protected against opportunistic behaviours; (3) gaining and ensuring legitimacy is achieved through the consent of affected parties and approval by governments; and, (4) achieving shock-absorption capabilities by the building of governability devices into the project structure; crisis funding, cohesion, reserves, flexibility, generativity and modularity.

References

Brown, S.L. and Eisenhardt, K.M. (1998) *Competing on the Edge: Strategy as Structured Chaos.* Boston, Harvard Business School Press.

Cleland, D. and Ireland, L. (2006) *Project Management: Strategic Design and Implementation, 5th edn.* New York, McGraw Hill.

Flyvbjerg, B., Bruzelius, N. and Rothengatter, W. (2005) *Megaprojects and Risk: An Anatomy of Ambition.* Cambridge, Cambridge University Press.

Herbert, A.S. (2000) Near decomposability and the speed of evolution. *Industrial and Corporate Change* **11** (3): 587–599.

Metcalf, J.S. (2005) *Innovation, Competition and Enterprise: Foundations for Economic Evolution.* Learning Economies Discussion Paper No. 71, CRIC University of Manchester.

Miller, R. and Lessard, D. (2001) *The strategic Management of Large Engineering Projects: Risks, Governance and Institutions.* Cambridge, MA, MIT Press.

Miller, R. and Lessard, D. (2007) *Client-driven Innovation in Large Engineering Projects.* Working Paper. Cambridge, MA, MIT Sloan School of Management.

Miller, R. and Olleros, X. (2007) The dynamics of games of innovation. *International Journal of Innovation Management* **11** (1): 37–64.

Mintzberg, H. (1994) *The Rise and Fall of Strategic Planning.* New York, Free Press.

11 Knowing differently, innovating together? Exploring the dynamics of knowledge creation across boundaries in clients' design teams

Patrick S. W. Fong

11.1. Introduction

Knowledge in designing a product does not form a complete and coherent body of knowledge that can be precisely documented or even articulated by a single individual. Rather, it is a form of knowing that exists only through the interaction among various collective actors (Gherardi and Nicolini, 2000). Nonaka (1994) has highlighted a need for the development of a diverse workforce if knowledge creation is to be promoted and sustained within organisations. This literature suggests that a diverse set of resources (experts with different backgrounds and abilities) provide a broad knowledge base at the individual level, offering greater potential for knowledge creation.

A multidisciplinary team is defined by Nonaka and Takeuchi (1995, p. 85) as 'a self-managed, self-organised team in which members from various functional departments, and/or areas of expertise, work together to accomplish a common goal'. The primary goal of the multidisciplinary composition is to marry diverse bodies of knowledge in a way that forces out a synergistic knowledge outcome that is innovative, contextualised, difficult to imitate and as such, has strategic value. For the most part, project team tasks are non-repetitive in nature and involve considerable application of knowledge, judgement and expertise.

The advantage of adopting multidisciplinary project teams is that they are quicker in integrating the expert knowledge of different functions, e.g. design, construction, property management, marketing, etc. If creating new collective knowledge is, indeed, a team level phenomenon, then the multidisciplinary team is considered the greenhouse where such a phenomenon can be best cultivated. In addition, a project on which a multidisciplinary team works, can metaphorically be seen as an experiment, a vehicle for knowledge creation with knowledge being created through the process of executing the project.

This research empirically investigates the creation of new technical knowledge in two multidisciplinary project teams, working on infra-structural and residential developments. It has important implications for those clients' project teams that are increasingly investing in knowledge-based co-operative ventures. As is the case with most construction projects, their raison d'être is creating knowledge to fulfil stake-holder needs, allowing for any constraints that may be imposed. Successful knowl-edge creation in the form of better design ultimately leads to economic benefits for the clients/customers that may result in the products meeting increasing demand. While the importance of new knowledge has been demonstrated and widely investigated (Brown and Eisenhardt, 1995), research that explores the endogenous knowledge cre-ation process in multidisciplinary project teams is in its infancy and offers fertile ground for study.

This chapter is structured as follows. After a brief introduction to the concept of knowledge creation, the literature on product development is considered. The method-ology adopted will then be discussed and the two case projects and their respective design work will be described and analysed. In the final section the overall findings will be discussed and conclusions from the research drawn out.

11.2. The importance of knowledge creation in new product development

Nonaka and Takeuchi (1995) have described new product development as a knowledge-intensive activity. A new product can be considered as 'a package of features and benefits, each of which must be conceived, articulated, designed and "operationalised", or brought into existence' (Dougherty, 1996, p. 425). The develop-ment of a constructed facility can be viewed as a new product development, with customers or end-users purchasing or using the facility. In addition, each project is unique in itself in terms of design and construction. With the many constraints the con-struction industry faces (due to limited space, increasing project complexity, limited budgets, tight programmes and the constant demand for clients driving innovation), project teams are faced with challenges to utilise diverse, and create new, knowledge to meet stringent requirements and fulfilling ever changing needs.

In this chapter, technological innovation and shared problem solving are viewed as knowledge creation activities. Leonard-Barton (1992) asserts that knowledge creation may involve problem solving, though problem solving need not involve knowledge creation. This is so because problems may be formulated and solved based on well-known knowledge without the need for creation, or even learning, to take place. Due to our common use of the term 'problem solving' in almost every situation, it is likely to downplay the tacit elements of knowledge creation and emphasise the explicit. Problem solving suggests that the necessary parameters of the 'problem' are known and the solution may be formed from determining the right combination of parameters. Thus new knowledge is created, or existing knowledge is combined, in those circumstances. Project design involves problem solving on many levels. A building project can be seen as a gap between an existing state and a desired state. The project delivery team members bring their experience and skills to bear on the problem of diminishing this gap. But design solutions are only a part of the design process, since permutation of team members, budgets, taste, priorities and values, for example, are also integrated. When taken together, the building appears as a collaborative effort created by, and within, the design process.

Technological innovation can be viewed as a form of problem solving (Dailey, 1978). That is, the solution to a problem is discovered after some amount of physical or mental exploration. Technological innovation as a result of knowledge creation in projects is claimed to be desperately needed in the construction industry (Gann, 2000). Viewing technological innovation and problem solving as knowledge-creating activities also follow the framework of Nonaka and Takeuchi (1995) in allowing for the concept of tacit knowledge. By acknowledging the tacit elements, the entire process of knowledge creation is included and may be more deeply understood (Anand *et al.*, 1993).

11.3. Methodology

The research described in this chapter is characterised as a comparative case study. The aim, to understand and conceptualise the interplay between the dynamics of knowledge creation and the development of boundary crossing, justifies the deployment of a case study method since it enriches our understanding of the context in conceptualising processes of knowledge creation and the impact of boundaries. Here, two project design teams – infrastructural (INF) and residential (RDA) – with diverse design concepts, discipline and knowledge bases, skills and possibly attitudes towards knowledge creation, are considered. Table 11.1 highlights various sources used for collecting evidence from both project teams. Both cases shared common involvement in the construction of two large-scale projects on a 'green field' site in Hong Kong. In addition, the nature of the work is information and knowledge intensive, requiring the teams to develop new or to utilise existing technologies, techniques and processes to achieve their work goals. However, these two projects differed in many respects. The nature of the tasks was different, as were the personnel involved (Table 11.2) and the ways in which design knowledge was created. These cases were selected in the hope of gaining further insight into the multiple and divergent phenomena fuelling the different modes of knowledge creation during design development.

The selection of the residential development project recognises the large reservoir of idiosyncratic knowledge developed by the property development company over the years. It also recognises the crucial innovating dynamics behind the need to compete on the market with other residential developments. The infrastructure project presented alternative opportunities for knowledge creation and learning, unique in several respects. Firstly, it was a complex operation, distinguished by an extraordinary multiplicity of consultants being employed. Secondly, it was rare to find such a project,

Table 11.1 Sources of evidence

Source of evidence	INF project	RDA project
Number of interviews	16	15
Average length of interview	80 min	100 min
Meetings attended (team meeting observation)	Formal – 12 Informal – 15	Formal – 16 Informal – 19
Access to company data	Open	Open
Access to project documentation	Open	Open
Informal discussions	Many	Many

Table 11.2   Professional services firms appointed for the INF and RDA projects

Professional service firms appointed for the INF project	Professional service firms appointed for the RDA project
Architectural	Architectural
Civil engineering	Building service engineering
Electrical and mechanical engineering	Environmental consulting
Environmental consulting	Interior design (clubhouse)
Landscape architecture	Interior design (floor and main entrance lobbies)
Pier consulting	Interior design (toilets and kitchens)
Quantity surveying	Lands consulting
Sewage treatment plant consulting	Landscape architecture
Submarine pipeline consulting	Structural engineering

usually managed by the government of Hong Kong Special Administrative Region, in private hands. Finally, the technical challenges presented in this project made it an interesting arena for knowledge creation and absorption within the team.

11.4. Case analysis

The term 'boundary crossing' refers to the process through which team members transcend various boundaries to share, generate, integrate and absorb relevant information and knowledge. Two types of boundary crossing from the case studies are found. Both existed at the outset of the two projects. The first boundary existed as a result of the specialist expertise of team members, initially defined by varying expertise expectations. The second boundary exists because of very distinct, hierarchical levels that lead to distancing and negative attributes. Typically, this boundary occurs at the interface of client, consultants and contractor. The discussion then highlights the processes used in crossing each boundary. Both boundaries were found to inhibit knowledge creation in multidisciplinary project teams.

11.4.1. Expertise boundaries within the project teams

Both the INF and RDA projects were complex, demanding input from different disciplines as no one professional was able to manage the whole development process. Both project team members appeared to possess distinctive knowledge, professional competence and specialist skills and seemed highly regarded by the general public. They generally co-operated fully with each other to ensure that the objectives set by the client were fully met.

The expertise boundary exists in project team members across disciplines. Expertise tended to be associated with specific individuals and labelled accordingly – for example, architect, structural engineer, quantity surveyor. Thus, an expertise boundary refers to whatever delimits the perimeter – and thereby the scope of a role. In both projects, the different expertise possessed by the team members resulted in knowledge gaps and created boundaries to build a shared understanding between team members. Some team members in both projects were accustomed to the traditional sequential design process, with one discipline completing a task before handing it over to another, with little team interaction or discussion aggravating the expertise division. Furthermore, the clear expertise barriers could create a design that was less than ideal, as each discipline pursued their own perspectives without due consideration for others. They might achieve optimal design in their own field but possibly not in terms of the overall, finished product. As illustrated in the INF project, the partial or complete absence of knowledge redundancy, due to a lack of familiarity with infra-structural design among team members, could create boundaries inhibiting knowledge development in a multidisciplinary team setting. This was due to the fact that the infrastructure project was funded and designed by a private property developer, whereas most installations would be under government jurisdiction. Previous work by Hutt *et al.* (1995) propose the concept of 'interpretive barriers' by referring to the difficulties created by participants' diverse knowledge backgrounds when making decisions. However, they do not proceed further to explain how the diversity of knowledge influences the creation of barriers.

With the presence of expertise boundaries, team members could dwell in their own disciplinary knowledge without due regard for other disciplines' needs and perspectives. Previous research finds that team members usually create and maintain boundaries as a means of simplifying and ordering the environment (Zerubavel, 1991). They erect 'mental fences' around people that appear to be similar, functionally related or otherwise associated. The boundaries are real in the sense that team members perceive them as such, acting as though they are real (Weick, 1979). These boundaries enable project team members to concentrate on whatever domain is currently salient to them, focusing less on other domains.

In the INF case, the submarine pipeline consultant could have designed the pipeline unilaterally, without looking into other team members' requirements. If this expertise boundary were maintained, he might achieve optimal design in his own field but possibly not beyond. In the RDA project, the same applied to the work between the environmental consultant and the architects or engineers. They could each go their own way but whether the building design could meet with the stakeholders' requirements would be seriously in doubt. In this project, some residential blocks were close to an existing railway line. Noise migration measures were of top priority in order to fulfil statutory requirements, especially environmental ones, as well as to minimise any adverse effects to future sales and the well-being of residents. Without crossing the expertise boundaries, the design could hardly satisfy the requirements of the stakeholders, including various regulatory authorities.

11.4.2. Observed methods for crossing expertise boundaries

The crossing of the expertise boundaries was in the form of connections with other team members. These could be the result of joint design activities, shared problem solving or knowledge redundancy among team members. Numerous examples could be found in both projects. Various INF disciplines were required to work together on the design of the pier and all the related team members were consulted. Besides looking at the structural integrity and at the impact of waves, the pier also needed to be atheistically pleasant as this would be a private pier for residents and visitors, to be funded by the developer but managed by government after completion. Other examples included the team of environmental consultants, architects and various engineers, in designing a sewage treatment plant that could fulfil statutory requirements for the different regulatory authorities. In the RDA project, resolving dimensional conflicts in the layout plans of different flats called for closer interaction among the architects, structural engineers and building services engineers as the issues could be intertwined.

In these particular case studies, some degree of knowledge redundancy between participants could have broken down the boundaries inhibiting cross-functional knowledge transfer. Team members suggested that knowledge overlap was achieved through prior knowledge acquired while working with other disciplines in previous projects, as well as experiences gained from working on similar facilities to the current ones. The INF team had fewer members with actual experience in infrastructure projects whereas most RDA team members had experience in previous residential design. In other words, there was little knowledge redundancy in terms of relevant experience amongst the INF participants. Under the normal infrastructure projects procured by government, consultants would be appointed separately to work on them rather than working interactively. The unique private nature of this infrastructure project had caused team members to lack the relevant project experience. In addition to possessing relevant project experience, it seemed advisable for team members to have some knowledge of professional disciplines other than their own. The research findings echo Nonaka's (1994) concept of knowledge redundancy. He explained that a degree of overlapping expertise between team members not only provides a platform to build shared understanding amongst them but also helps them acquire new knowledge.

In both cases, there was other evidence of crossing the expertise boundaries. The catalyst was often a technical hurdle requiring multi-disciplinary problem solving. The interaction sometimes began when a team member risked stepping beyond his own role and boundaries to challenge the work or logic of someone else. In the INF project, boundary crossing, or stepping into another person's territory, was necessary when shared problem solving, specific to infrastructure design, was required. For example, the sewage treatment plant design specifically called for a high level of collaboration among various team members; besides technical performance, siting location and how to blend in with the surrounding environment, multidisciplinary professional skills and judgement were also required. Boundary crossing was observed within the RDA team, when a team member offered insight previously gained from within a colleague's knowledge domain. This facilitated the crossing of the expertise boundaries.

When people develop a level of comfort within a group setting, they are able to question each other without anyone becoming defensive. During informal meetings, it became evident that team interactions were relaxed and spontaneous, characterised by little pretence. Such environments were conducive to transcending boundaries and entering other people's disciplines. The basis of interaction shifts from an evaluation of another perspective to valuing and exploring multiple perspectives. This allows other ideas to expand one's own understanding. Knowledge creation in this mode requires time in which to discuss issues, along with patience to hear and fully appreciate another person's perspective. An example of this could be seen when the project teams faced technical hurdles. Both teams would collect as many facts about the problem as possible and through discussion; they would aim to identify the core issue, be it technical, aesthetic or regulatory. Through such crossing of individual expertise boundaries, they were able to offer and receive insights from other team members and resolve many key issues.

Identifying core problem areas and knowledge gaps is a primary advantage of this level of cross-disciplinary integration. Considerable time and resources can be wasted if team members pursue peripheral notions. Team members in both projects had opportunities to interact across expertise boundaries during formal and informal meetings. They were able also to communicate the information and knowledge needed for the projects through the use of boundary objects. Boundary objects are physical artefacts that enable people to understand others' perspectives. They help people to cross boundaries and come together to solve a problem by inhabiting 'several intersecting social worlds' and satisfying 'the information requirements of each of them' (Star and Griesemer, 1989, p. 393). As observed within the project setting, boundary objects could be concrete or abstract, including sketches and drawings, reports, correspondence, tender documents or specifications. They would have to be tangible to all team members, with the capacity to support translation across boundaries. Generally, concrete objects are a pivotal feature of crossing disciplines, often shared between different professions and sometimes brought together for discussion when requiring input from other disciplines. Boundary objects may also be critical in highly interactive situations, as in both case studies. The project teams accomplished important boundary crossing through team conversations with the aid of boundary objects. The objects had different meanings within different disciplines but shared a common structure, making them recognisable to all team members. As a result of their flexible structure, they served as a means of translation between the various disciplines.

Participants used drawings as a primary means of communicating, often pulling documents out in the course of conversation. Alternatively, they might have prepared free-hand sketches during a discussion to support their case. Since creating and interpreting drawings was an important aspect of the project team's work, they shared a common understanding through this boundary object. Evidence from both cases confirmed that boundary objects are representations or artefacts that embody particular aspects of knowledge-in-practice, yet have a shared or understandable character across different discipline settings, making them useful for mutual or cross-functional problem-solving.

11.4.3. Hierarchical boundaries within the project teams

The two case studies' findings suggest that hierarchical differences could easily become boundaries separating participants in the knowledge creation process from non-participants. The implication for cross-functional knowledge creation is that hierarchical boundaries create barriers that restrict communication as well as various knowledge-creating activities. Manifest in both case studies, the difficulties created by hierarchical differences are not necessarily related only to the willingness of team members to share their knowledge with others. Just as important is the issue of whether a shared understanding can be achieved among team members.

The hierarchical differences, although seldom voiced openly, did affect behaviour in boundary crossing, with most team members consciously drawing distinction between the client, consultant and contractor. The interview process revealed that these types of hierarchical boundaries did exist in both project teams. Some team members may have been reluctant to openly share ideas with the client, possibly fearful of appearing unintelligent or assuming that the client, as an experienced developer, probably had all the answers regarding facilities development. Consultants may have been reluctant to acknowledge the participation of contractors during the design process, as they are presumed to be only concerned with ease of construction without due regard for aesthetic considerations. In addition, they may perceive the architects, or the architects perceive themselves, as the lead consultants on the projects, with their contributions valued higher than other team members. This, in fact, could be the consequence of the construction industry's current procurement practice of having a clear divide between design and construction, as well as a distinction between the different professional statuses of consultants. As Argyris (1995) demonstrates in his research, these attributions typically go untested. The behaviour of participants in any given situation is based on their beliefs. The fact that they hold particular beliefs and assumptions is not open for discussion. This self-fulfilling sequence erected an even higher boundary in both project teams between client, consultants and contractor. Because of the positional power of client versus consultant versus contractor in the decision making process, high and low status members were identified and acknowledged in both teams.

11.4.4. Observed methods for crossing hierarchical boundaries

Top-down boundary crossing means that the client's project managers have a need to gain the full collaboration of the consultants and contractors involved in the project teams. The negative impact of these distinct boundaries lies in their potential damage to knowledge-creating opportunities among project team members, possibly limiting their contributions when they have so much to offer. In meetings, it was evident, through the dialogue between the consultants and the contractor team, that the team members began to recognise and acknowledge their construction expertise in related areas as the project manager emphasised the changing company culture regarding design and construction. Examples in the INF case included the joint design effort among the consultants and contractor for the system formwork and the concrete retaining wall next to a proposed school site for the project. The consultants were also reassured that problems during construction, possibly created by a complicated design, could be resolved at a much earlier stage. All the while, the project managers served to reinforce the importance of removing hierarchical boundaries. Inviting the contractors to participate in formal meetings meant that construction knowledge could be accessed constantly. It also acknowledged their expertise and served to lower the boundary barriers. The RDA project managers had unceasingly encouraged project team members with any contributions to the project to speak out, even if outside their professional boundaries. Their constant invitation in having new ideas from team members reflected their open attitudes towards ideas that could make the project distinctive as well as lessen the effect of professional segregation.

Table 11.3 focuses on illustrating the contrast or similarity between various features and activities that occurred within the two different clients' design teams.

Table 11.3 Key comparisons between the INF and RDA project teams

	INF client's project team	RDA client's project team
Diversity	Diverse team of professionals	Diverse team of professionals
Past experience	Not many team members had direct and relevant experience	Most of them possessed relevant previous experience
Client type	Open-minded private client with little experience in infrastructure development	Innovative and competitive private client with ample experience in residential development
Boundaries		
• Expertise boundaries	Expertise boundaries existed among client's project team members	Expertise boundaries existed among client's project team members
How to cross:	Multi-stakeholders' requirements demanded in the product; design challenges *or* technical hurdles; use informal meetings; facilitated by the use of boundary objects	Intertwined design inputs to product; prior project experience; offering insights in other's knowledge domain; use informal meetings; facilitated by the use of boundary objects
• Hierarchical boundaries	Perceived hierarchical boundaries existed between client, consultant and contractor	High-low status existed among project participants, creating hierarchical boundaries among client's project team members
How to cross:	Joint design effort; top-down boundary crossing implicated by the client; active participation of client's project managers	Deliberate invitation by the client to share ideas; active participation of client's project managers

11.5. Conclusions

The study describes two types of boundaries affecting the progress and success of multidisciplinary knowledge creation in both projects. The importance of boundary crossing is reflected in the notion of 'solving the boundary paradox' (Quintas *et al.*, 1997) where team members are able to exchange and combine knowledge. The interpersonal interactions across these boundaries can either foster or hinder knowledge creation. In considering the case studies, the first boundary was between team members of different disciplines. The second boundary existed between client, consultant and contractor.

As described above, the expertise boundaries could be crossed not only through shared problem solving and knowledge redundancy among team members, but also through boundary objects. The most prominent project boundary objects were drawings and personal conversations among team members. The second hierarchical boundary could be crossed by the conscious effort of team members to break down barriers by valuing the expertise of others. The example set by the clients' project managers was also helpful in this regard.

References

Anand, P., Choi, C., Grint, K. and Hilton, B. (1993) *Knowledge, Understanding and Trust in Deontologic, Teleologic and Epistemic Organisations.* Management Research Paper, Templeton College, Oxford Centre for Management Studies.

Argyris, C. (1995) Action science and organizational learning. *Journal of Managerial Psychology* **10** (6): 20–26.

Brown, S.L. and Eisenhardt, K.M. (1995) Product development: past research, present findings, and future directions. *Academy of Management Review* **20** (2): 343–378.

Dailey, R.C. (1978) The role of team and task characteristics in R&D team collaboration problem solving and productivity. *Management Science* **24** (15): 1579–1589.

Dougherty, D. (1996) Organizing for innovation. In: *Handbook of Organization Studies* (eds S.R. Clegg, C. Hardy and W.R. Nord). London, Sage, pp. 424–439.

Gann, D. (2000) *Building Innovation: Complex Constructs in a Changing World.* London, Thomas Telford.

Gherardi, S. and Nicolini, D. (2000) The organizational learning of safety in communities of practice. *Journal of Management Inquiry* **9** (1): 7–18.

Hutt, M., Walker, B. and Frankwick, G. (1995) Hurdle the cross-functional barriers to strategic change. *Sloan Management Review* **36** (3): 22–30.

Leonard-Barton, D. (1992) The factory as a learning laboratory. *Sloan Management Review* **34** (1): 23–38.

Nonaka, I. (1994) A dynamic theory of organizational knowledge creation. *Organization Science* **5** (1): 14–37.

Nonaka, I. and Takeuchi, H. (1995) *The Knowledge-Creating Company: How Japanese Companies Create the Dynamics of Innovation.* Oxford, Oxford University Press.

Quintas, P., Lefrere, P. and Jones, G. (1997) Knowledge management: a strategic agenda. *Long Range Planning* **30** (3): 385–391.

Star, S.L. and Griesemer, J.R. (1989) Institutional ecology, 'translations' and boundary objects: amateurs and professionals in Berkeley's Museum of Vertebrate Zoology, 1907–39. *Social Studies of Science* **19**: 387–420.

Weick, K.E. (1979) *The Social Psychology of Organizing.* Reading, MA, Addison-Wesley.

Zerubavel, E. (1991) *The Fine Line: Making Distinctions in Everyday Life.* New York, Free Press.

12 The role of the client in the innovation processes of small construction professional service firms

Shu-Ling Lu

12.1. Introduction

Knowledge and knowledge workers are increasingly being regarded as the key source of wealth for individual organisations and nations. Small construction professional service firms (PSFs) form a core part of the construction industry and, therefore, are important actors in any endeavour to enhance overall construction industry performance. PSFs are characterised by firms whose service offerings are client specific and have, at their core, the generation and application of new knowledge (Alvesson, 2001), and which are co-produced by the professional and the client (Løwendahl, 2000). There is significant agreement that the principal means by which this growing body of PSFs create value is through the successful creation and management of knowledge (e.g. Robertson *et al.*, 2001). The 'value-creating' performance of the construction industry, however, has often been questioned by its clients (DETR, 1998). The common perception of the construction industry is that of an industry that delivers products and services which are often of inappropriate quality, and which fail to meet client demands for price certainty and guaranteed delivery. As a consequence, the client is the key in improving the construction performance. The focus of this chapter is on the role of the client in the innovation process within small construction PSFs (see Lu and Sexton, 2006 for a fuller discussion). This chapter is presented as follows. First, key issues from the innovation literature in construction are discussed. Second, a case study methodology deployed in this research is described. Third, key findings from the case study are reported. Finally, discussion and conclusions are drawn.

12.2. Key issues from the innovation literature

12.2.1. Knowledge-based view of innovation

There is a diverse range of definitions of innovation in the literature, but a recurring theme across the definitional debate is that 'new ideas' are taken to be the starting point for innovation (e.g. Rogers, 1983). The emphasis on newness is prominent in the construction literature (e.g. Sexton and Barrett, 2003a, 2003b). The pertinent issue for small construction PSFs is that the 'new ideas' are intrinsically 'knowledge-laden' and are principally the outcome of the co-production between the knowledge worker and the client. Hansson (2002), for example, states that the generation of successful services demands a high degree of interaction and co-production of the service provision between the client and the service provider. Hill and Neely (1988), for instance, characterise a 'professional service' as one where the client is significantly dependent on the provider to define the problem and give appropriate advice. In this context, the client is considered as a key actor in the innovation processes and performance in such firms.

The proposition here is that the development of the optimal dynamic capabilities which bring the client and the knowledge worker together to co-produce innovation is the principal source of sustainable competitive advantage for the small construction PSF. Innovation for small construction PSFs should, therefore, be considered synonymous with a 'knowledge-based' view of innovation which consists of knowledge-based resources and capabilities, labelled here as 'knowledge capital'. This 'knowledge capital' is presented as the 'dynamic innovation capability', which generates innovation and sustainable competitive advantage within small construction PSFs. The following sections concentrate on what kinds of knowledge-based resources and capabilities are required to create, manage and exploit innovation within a SCPSF context.

12.2.2. Knowledge-based resources for innovation

The identification of different types of knowledge created within organisations is the first step to understanding how successful knowledge-based innovation is brought about. Knowledge has been traditionally grouped into two complementary types: tacit and explicit (Polanyi, 1962, 1967). The tacit and explicit distinction has evolved into knowledge as a 'noun', i.e. an 'asset' which can be neutrally articulated, stored and traded (explicit knowledge); and, knowledge as a 'verb', i.e. the context specific 'process' of knowledge creation and use (tacit knowledge). De Long and Fahey (2000) synthesise fruitfully the 'asset' and 'process' dimensions, and identify three distinct, but interactive, types of knowledge:

(1) Human knowledge constitutes what individuals know or know how to do, and is manifest in experience, knowledge and skills. Human knowledge is tacit knowledge.
(2) Relationship/Social knowledge exists in relationships among individuals and groups, which add value to activities. Relationship knowledge is largely tacit, composed of cultural norms that exist as a result of working together. Relationship knowledge is reflected by an ability to create and maintain effective collaboration.
(3) Structure knowledge is embedded in organisational systems, processes, tools, rules and routines. Structure knowledge is largely explicit and rule-based and can exist independently of staff.

These three types of knowledge are critical to understanding innovation in small construction PSFs. The argument here is that the appropriate generation of, and conversion between, human knowledge, relationship knowledge, and structure knowledge, is essential to successful knowledge creation and thus (particularly in small construction

Table 12.1   Knowledge-based innovation concept model variables

Variables	Brief description
Interaction environment	It is that part of the business environment which firms can interact with, and influence, including the 'task environment' (the environment where this client interaction occurs) and the 'competitive environment' (the environment where firms compete for customers and scarce resources). The interaction environment and the firm are separated by an organisational boundary
RC	It is the network resource of a firm. It results from interactions between individual, organisation, and external supplier chain partners
HC	It is defined as the capabilities and motivation of individuals within the firm and external supply chain partners to perform productive, professional work in a wide variety of situations
SC	It is made up of systems and processes (such as company strategies, computers, tools, work routines and administrative systems) for codifying and storing knowledge from individual, organisation and external supply chain partners
KC	It is the dynamic synthesis of both the 'context' and 'process' of knowledge creation and conversion within IOI knowledge ba spirals, and the 'content' of RC, SC and HC

PSFs) successful innovation. It is proposed that human knowledge, relationship knowledge, and structure knowledge are embedded in human capital (HC), relationship capital (RC) and structure capital (SC), and these resources make up knowledge capital (KC) (Each resource is described in Table 12.1). The rationale for the capabilities required by small construction PSFs to produce KC is explored below.

12.2.3. Organisational capabilities for innovation

The organisational capability for innovation is defined, for example, as 'the comprehensive set of characteristics of an organisation that facilitate and support innovation strategies' (Burgelman *et al.*, 1996, p. 8). It has been argued that the acquisition of 'organisational capability' occurs through 'organisational learning' processes (Chaston *et al.*, 1999) and that 'organisational learning' leads to innovation (Argyris and Schön, 1996), particularly in small and medium sized enterprises (Chaston *et al.*, 1999). A key challenge for companies is when to innovate and when not to innovate. The work of March (1991, p. 65) provides theoretical guidance on this challenge through the distinction between exploitative and explorative routines, and the need for appropriate balance between them. They are adapted and defined as follows:

(1)   Exploitative capabilities are those resources utilised to improve organisational *efficiency* to generate *short-term* competitive advantage.

(2)　Explorative capabilities create and use new resources and capabilities to improve organisational *effectiveness* to generate *sustainable* competitive advantage.

The concepts of exploitative and explorative capabilities are a useful way of understanding, connecting and managing knowledge-based resources for successful innovation. This proposition leads to the concept of successful knowledge-based innovation (see 'knowledge-based innovation concept model' section below). The next section will address the key managerial challenges facing small construction PSFs in developing and using these knowledge-based capabilities.

12.2.4.　Key managerial challenges for innovation

It is argued that RC plays a particularly important role in innovation (Ibarra, 1993). Clients and their networks, as well as professional networks, are important resources for PSFs (Løwendahl, 2000). Second, appropriate HC is the essential factor to bundle different resources and capabilities to form KC to bring about appropriate innovation in services and service delivery (Maister, 1993). Baumard (2002), for example, argues that the generation and stimulus for 'new ideas' requires the motivation and in-depth knowledge and experience of knowledge workers. For small construction PSFs, RC is the principal stimulus for innovation and, therefore, presents key challenges for successful innovation.

　　To reiterate, the co-production of professional services demands a high degree of interaction between knowledge workers and clients. As a consequence, client involvement in the innovation process is vital for driving innovation in small construction PSFs. Knowledge sharing and creation is thus significantly based on HC held by knowledge workers within PSFs (e.g. Sverlinger, 2000). Knowledge workers' knowledge about customers tends to be personal and anecdotal, and situationally prescribed (Clippinger, 1995, p. 28). This 'person specific' knowledge held by knowledge workers is labelled as 'individual knowledge' (Simon, 1957). The accrued or cumulative learning and knowledge of individuals has been referred to as 'individual knowledge capital' (Neilson, 1997, p. 1). The challenge within small construction PSFs is to combine various individual knowledge domains to form dynamic 'organisational knowledge' in new configurations, which feed back into, and enrich, individual knowledge (Bhatt, 2002). Knowledge within the organisational level forms organisational KC. The proposition made here is that organisational KC within small construction PSFs arises from a dynamic spiralling process wherein RC, SC and HC are converted into relationship knowledge, structure knowledge, and human knowledge through their exploitative and explorative capabilities. Hence, these constant 'interaction activities' form an individual–organisational–individual (IOI) KC spiral. Through this spiral, individual KC is converted into fresh organisational KC and allows other individuals to access the organisational KC base. As a consequence, KC is dynamic (explorative capability), but must be capable of being accessed and used at any given time (exploitative capability).

　　Nonaka *et al.* (2001) further indicate that these 'interaction activities' take place in the 'ba', which is a place, space or facility where individuals interact to exchange ideas, share knowledge, conceptualise and create new knowledge. Davenport and Prusak (1998, p. 137) argue that the 'ba' should be focused on the 'knowledge' environment. 'Ba' is thus labelled as 'knowledge ba'. For small construction PSFs, the 'knowledge

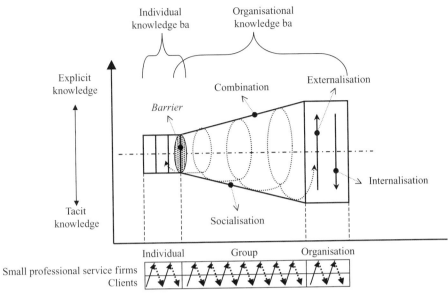

Figure 12.1   Spiral of organisational knowledge capital.

ba' is predominantly located within the interaction between individual knowledge workers and their clients. This individual level of the 'ba' can be viewed as 'individual knowledge ba'. There is a need, therefore, for a shared context for knowledge creation and conversion to take place from the 'individual level' to 'organisational level', and then back to the 'individual level' (Nonaka and Takeuchi, 1995). The organisational level of the shared context can be viewed as 'organisational knowledge ba'.

Insights into the nature and process of the interaction between individual and organisational knowledge ba can be gleaned from the adoption of the knowledge spiral model presented by Nonaka and Takeuchi (1995, p. 73). Figure 12.1 shows different phases of knowledge interactions between clients and the SCPSF, including individual, group and organisational interactions. First, knowledge interactions start at the individual level. This is shown in Figure 12.1 in the left hand side of the bottom rectangle. Knowledge interactions at the individual level occur in the 'individual knowledge ba'. Second, knowledge dialogue expands outside individual knowledge worker–client interactions. At this stage, the collaborative interactions of individuals share their diverse interests and issues within a team context. As the knowledge work tends to be project-based and extends to the broader organisation and beyond, the knowledge interaction occurs across 'individual' and 'organisational' knowledge ba. Organisational knowledge ba thus presents an influential factor facilitating the IOI knowledge creation and conversion spiral within small construction PSFs. This spiral, which continuously nurtures the interaction and development of individual and organisational knowledge ba, is taken to be the core dynamic innovation capability for small construction PSFs. This argument leads to the concept of KC (see 'knowledge-based innovation concept model' section below). From this discussion, the managerial challenge of 'how do small construction PSFs appropriately develop and manage knowledge interaction activities within IOI knowledge ba spirals, and how do these arrangements affect innovation performance?' appears significant.

12.3. Case study methodology

An interpretative philosophy was adopted for this research. The rationale for this is that the author adopts the view that innovation in PSFs cannot be reduced to rational cause and effect relationships; rather, it is a product of idiosyncratic social constructions. Further, the motivation of the knowledge worker requires individual interpretations of the consequence of specific behaviour and therefore cannot be brought together in unconditional causal generalisations that enable the researcher to predict and control individual human actions (Rosenberg, 1994). Within this context, a 22 month single case study research approach was used with an exploratory phase and an action research phase. The research techniques for secondary data collection consisted of a review of the relevant literature; and, for primary data, semi-structured interviews, company documentation, action research 'real world' activities and workshops. The primary data analysis research techniques comprised content analysis and cognitive mapping.

12.4. Key research results

This section presents the key results from the case study. This section is organised as follows. First, background of the case study company is introduced. Second, the definition of an appropriate successful knowledge-based innovation and a concept model of knowledge-based innovation are described. Third, two types of knowledge-based innovation are introduced. Fourth, a definition of a successful knowledge-based innovating firm is given. Finally, key innovation management challenges are set out.

12.4.1. Background of the case study company

The case study company, labelled hereafter as ArchSME for confidentiality reasons, is an architectural design practice located in Manchester in the northwest region of England. Its principal markets are the residential sector: varying from one-off commission from domestic clients to repeat business from national house builders. A small architectural design practice was chosen for the case study for two reasons: the practice was of an appropriate 'small' size, having 40 staff (e.g. EC, 2003, p. 39); and, the practice was of the 'archetype' of a PSF (Day and Barksdale, 1992; Wilson, 1997). The case study firm is a recent start-up with three equity directors within a limited liability partnership structure. The firm has grown from an initial turnover of £0.3 million in 1999 to £2 million in 2004. In the same period, the firm has increased its staff numbers from 12 to 40.

12.4.2. Knowledge-based innovation concept model

The research findings defined appropriate knowledge-based innovation as: 'the effective generation and implementation of a new idea which enhances overall organisational performance, through appropriate exploitative and explorative knowledge capital which develops and integrates relationship capital, structure capital and human capital'. This definition forms a knowledge-based innovation concept model

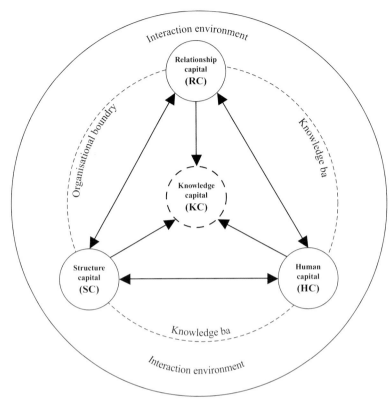

Figure 12.2   Knowledge-based innovation concept model for SCKIPSFs.

(Figure 12.2), which is built around four variables that are summarised in Table 12.1. The concept model highlights the need for senior management to strategically and systemically build, connect and energise appropriate RC, HC and SC to form KC, from which successful organisation and project innovation will flow.

12.4.3. Two forms of knowledge-based innovation

Two types of knowledge-based innovation were distinguished: explorative innovation and exploitative innovation (Figure 12.3). 'Explorative innovation' focused on client facing, project-specific problem solving, often in close collaboration with the client. Explorative innovation activity heavily relied on the capacity, ability and motivation of staff at an 'operational level' to solve client problems to generate short-term competitive advantage (i.e. project specific). Their outcomes focused on effective and efficient delivery of services to satisfy prevailing fee-earning project needs, but were often not embedded in the organisational SC due to management attention and company resources being constantly focused on other current or near future project-specific demands. The use of new materials in the case study firm, for example, was explorative in nature, being project-specific and individually driven. In contrast, 'exploitative innovation' focused predominantly on internal organisation and general client development activity

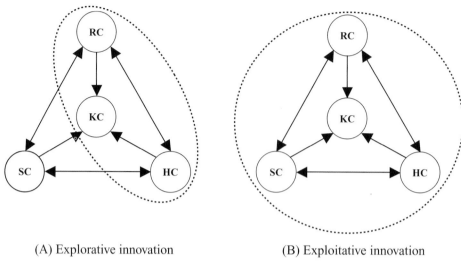

(A) Explorative innovation   (B) Exploitative innovation

Figure 12.3   Types of knowledge-based innovation.

(non-project-specific, fee-earning activity). The client had no direct involvement with the innovation activity. Exploitative innovation activity heavily relied on the capacity, ability and motivation of senior management at a 'social' level to improve organisational effectiveness and efficiency to generate sustainable competitive advantage.

The distinctive feature of exploitative innovation (compared to explorative innovation) was that new phenomena, systems or structures were securely embedded in the SC of the firm. The motivation for a new mission statement for the case study firm, for example, came from senior management, who saw it as a way of instilling an integrating vision for its portfolio of activities. Key generic and distinctive variables around these two types of innovation are summarised in Table 12.2. The Table shows that the key distinction between successful and unsuccessful innovation was the 'social' or 'operational' knowledge being applied to a specific innovation. 'Operational knowledge' was generated and created in 'operational level' interactions where the focus was on solving project-specific issues/problems. These projects were either 'external' fee-earning projects, or 'internal', but specific client-driven projects. 'Social knowledge' was generated through 'social level' interactions where the focus was on generating non-project-specific innovation which built up general organisational capability, and forged and replenished deeper client relationship over the medium to long term. Moreover, social knowledge was found to have a significant effect on feeding operational knowledge at a specific project level at a future date.

12.4.4. Definition of a successful knowledge-based innovating firm

The research findings reveal that successful explorative innovation did not necessarily need integrated SC. It was evident in the case study firm that there was too much emphasis on individual learning at the project level (explorative innovation) to the detriment of the organisational level learning (exploitative innovation). This emphasis of explorative KC over exploitative KC is not sustainable as the limitation of SC will

Table 12.2 Key generic and distinctive variables for explorative and exploitative innovations

Type of innovation	Variables	Generic variables	Distinctive variables for successful innovation	Distinctive variables for unsuccessful innovation
Explorative innovation	HC	• The capacity, ability and motivation of staff	• Social and operational knowledge being applied to meet project needs	• Social knowledge not being applied to meet project needs
	SC	• Team structure • Teamwork	• Team-based ideas • Teamwork • Senior management involvement through teamwork	• Individual-based ideas • Individual-based work • Senior management not involved in teamwork • Limitation of relevant and updated information within the structure
	RC	• Operational RC: within internal, client and supplier interactions • Social RC: within internal, client and supplier interactions	• Operational and social RCs	• Social RC
	KC	• Social context[a]: company environments (office, meeting room) • Technical context[b]: e-mails, internet	• A combination of social context and technical context being applied to meet project needs	• Technical context
	Outcome	• Effective and efficient delivery of services to satisfy current and/or future project needs	• Project performance improvement	• Individual performance improvement
Exploitative innovation	HC	• The capacity, ability and motivation of senior management • Employee participation	• Top management support • Senior management implementation • Employees buy into the need for innovation through training	• Top management not supportive • Senior management not driving the implementation • Lack of time

(cont.)

Table 12.2 *(cont.)*

Type of innovation	Variables	Generic variables	Distinctive variables for successful innovation	Distinctive variables for unsuccessful innovation
	SC	• The administrative system • Team structure • Computer systems	• Formalised structures and documentation systems • Senior management implementation through the team structure	• Employees not buy into the need for innovation (inappropriate encouragement and not related to an individual job) • No formalised structures and documentation systems • Senior management not driving the implementation through the team structure
	RC	• Operational RC: within business adviser, internal, client and supplier interactions • Social RC: within internal interactions	• Operational and social RCs	• Social RC
	KC	• Social context[a]: company environments (office, open family culture) • Technical context[b]: e-mails, internet	• A combination of social context and technical context being applied to meet project needs	• A combination of social context and technical context not being applied to meet project needs
	Outcome	• Organisational effectiveness and efficiency	• Organisational performance improvement	• Individual performance improvement

[a] A 'social' context was used to stimulate interaction and collective 'process orientated' knowledge creation and conversion.
[b] A 'technical' context was used to support the search for external knowledge and sharing of 'asset orientated' knowledge.

become increasingly evident as a significant restraining force for the effective integration of explorative and exploitative KCs. There was thus not an appropriate balance between explorative and exploitative innovation over time. The case study firm needed to create and continue a balance between explorative and exploitative KC barriers, which would allow the flow of KC between operational and social levels. The following definition of a successful knowledge-based innovating firm is therefore offered to accommodate the time dimension: 'the effective generation and implementation of a flow of new ideas which enhance overall organisational performance *over time*, through appropriate exploitative and explorative knowledge capital which develops and integrates relationship capital, structure capital and human capital'.

The time variable brings into focus the development phases of firms as they move from start-up to mature organisations. The focus and process of innovation activity will correspondingly change during the transition. It can be speculated that at the early stages of firms' development, the emphasis is on explorative innovation, but, as firms matures, there is an increasing need to explicitly invest in exploitative innovation. This need was certainly evident in the case study firm.

12.4.5. Key innovation management challenges

The discussion above has provided insight into the nature and process of innovation for small construction PSFs. It was achieved by answering the research question set out in the 'key managerial challenge for innovation' section above. The answer to the question is that successful innovation in small construction PSFs is principally characterised by 'project pull' and 'project push' IOI knowledge ba spirals which create dynamic specific-project and/or client-driven KC. Figure 12.4(A) depicts specific project requirements (either external fee-producing or internal client-driven projects) 'pulling', combining and converting, 'organisational knowledge' and 'individual knowledge' to form specific 'project individual knowledge'. Project individual knowledge is integrated and leveraged to create 'project team knowledge' which is appropriately applied to create successful innovation. The feedback IOI knowledge ba spiral is complemented by a feedback or 'project push' knowledge ba spiral where new specific 'project team knowledge' feeds back to develop 'project individual knowledge', which, in turn, further enhances 'individual knowledge' and 'organisational knowledge'. The tacit, experiential knowledge accumulation and learning is the basis for subsequent cycles of project-based innovation. In contrast, unsuccessful innovation in small construction PSFs is principally characterised by 'organisation push' of disjointed, unfocused 'social' non-project-specific and/or non-client-driven KC being 'rejected' by day-to-day project priorities and activities. Without a specific-project focus, innovation fails because the IOI knowledge ba spiral does not happen.

Figure 12.4(B) depicts that there is no specific project needs 'pulling' 'individual knowledge' and 'organisational knowledge' together. Rather, generic 'organisational knowledge' is 'pushed' into a project team setting without appropriate filtering and adaptation to meet specific project needs. Further, the 'organisational knowledge' does not benefit from individual knowledge worker championing and tacit understanding. In combination, the 'organisational knowledge' is 'rejected' by day-to-day projects. As a consequence, the feedback loop through individual knowledge, project knowledge and organisational knowledge does not happen.

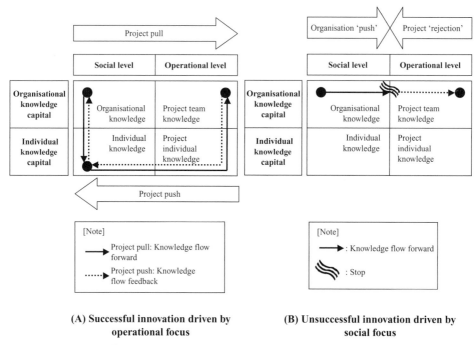

(A) **Successful innovation driven by operational focus**

(B) **Unsuccessful innovation driven by social focus**

Figure 12.4   The difference focus in driven successful and unsuccessful innovations.

12.5. Discussion and conclusions

The research results confirm that RC provides a critical network of contacts to enable creative action. This is consistent with the literature that RC provides access to knowledge-based resources and is a valuable source of information (e.g. Hendry *et al.*, 1995). Baker (2000), for example, argues that it is not 'what you know', but 'whom you know'. In addition, the research findings reveal 'clients' as being the principal agent in the interaction environment (see Table 12.1 for its definition). It was evident that the initial ideas for explorative innovation were to meet specific project needs (client needs); and, the initial ideas for exploitative innovation targeted client-driven business needs. This is consistent with the literature by Schneider and Bowen (1995), who argue that service productivity is, to a significant degree, influenced by the exchange of information and resources between the service provider and client. The importance of client relationships view is emphasised by Tapscott *et al.* (2000, p. 12), who argue that 'the wealth embedded in customer relationships is now more important than the capital contained in land, factories, buildings, and even back accounts'.

This chapter has investigated the role of clients in the innovation process within small construction PSFs. The research findings emphasise the overtly social nature of innovation in small construction PSFs: it is not a mechanistic, linear process; rather, it is a fluid process where knowledge-based innovation flows from context specific 'one-off' encounters between clients and the knowledge workers at the project level of resolution. The success of small construction PSFs in the construction industry is heavily dependent on the creation and exploitation of innovation within and through localised client–professional interaction, which appropriately responds to client and project specific characteristics and needs. The case study results further indicate that the client has a direct role in driving explorative innovation; and, has an indirect role in driving exploitative innovation in small construction PSFs. As innovation activity is explorative and exploitative in nature, the key challenge for small construction PSFs is to develop and manage an appropriate balance between explorative and exploitative innovation over time in order to generate sustainable competitive advantage.

References

Alvesson, M. (2001) Social identity and the problem of loyalty in knowledge-intensive companies. In: *Knowledge Work, Organisations and Expertise: European Perspectives* (eds F. Blackler, D. Courpasson and B. Elkjaer). London, Routledge, pp. 582–587.

Argyris, C. and Schön, D. (1996) *Organization Learning II: Theory, Method and Practice*. Reading, MA, Addison-Wesley.

Baumard, P. (2002) Tacit knowledge in professional firms: the teachings of firms in very puzzling situations. *Journal of Knowledge Management* **6** (2): 135–151.

Baker, W. (2000) *Achieving Success Through Social Capital: Tapping the Hidden Resources in Your Personal and Business Networks*. San Francisco, CA, Josey-Bass.

Bhatt, G.D. (2002) Management strategies for individual knowledge and organizational knowledge. *Journal of Knowledge Management* **6** (1): 31–39.

Burgelman, R., Maidique, M. and Wheelwright, S. (1996) *Strategic Management of Technology and Innovation*, 2nd edn. Homewood, Irwin.

Chaston, I., Badger, B. and Sadler-Smith, E. (1999) Organizational learning: research issues and application in SME sector firms. *International Journal of Entrepreneurial Behaviour and Research* **5** (4): 191–203.

Clippinger, J.H. (1995) Visualization of knowledge: building and using intangible assets digitally. *Planning Review* **23** (6): 28–31.

Davenport, T.H. and Prusak, L. (1998) *Working Knowledge: How Organizations Manage What They Know*. Boston, Harvard University Press.

Day, E. and Barksdale, H.C. (1992) How firms select professional services. *Industrial Marketing Management* **21**: 85–91.

De Long, D.W. and Fahey, L. (2000) Diagnosing cultural barriers to knowledge management. *Academy of Management Executive* **14** (4): 113–127.

Department of the Environment, Transport and Regions (DETR) (1998) *Construction Statistics Annual: 1998 Edition*. London, DETR.

European Commission (EC) (2003) Commission recommendation of 6 May 2003 concerning the definition of micro, small and medium-sized enterprises. *Official Journal of the European Union* **L124**: 36–41.

Hansson, J. (2002) *Management of Knowledge Transfer in Knowledge Service Firms*, Paper for EURAM 2002: Innovative Research in Management, 9–11 May, Stockholm, Sweden.

Hendry, C., Arthur, M. and Jones, A. (1995) *Strategy Through People: Adaptation and Learning in the Small-Medium Enterprise*. London, Routledge.

Hill, C.J., and Neely, S.E. (1988) Differences in the consumer decision process for professional vs. generic services. *Journal of Services Marketing* **2** (1): 17–23.

Ibarra, H. (1993) Network centrality, power, and innovation involvement: determinants of technical and administrative roles. *Academy of Management Journal* **36** (3): 471–501.

Løwendahl, B.R. (2000) *Strategic Management of Professional Service Firms*, 2nd edn. Copenhagen, Handeshøjskolens Forlag.

Lu, S. and Sexton, M. (2006) Innovation in small construction knowledge-intensive professional service firms: a case study of an architectural practice. *Construction Management and Economics* **24**: 1269–1282.

Maister, D.H. (1993) *Managing the Professional Service Firm.* New York, Simon and Schuster.

March, J.G. (1991) Exploration and exploitation in organizational learning. *Organization Science* **2** (1): 71–87.

Neilson, R.E. (1997) *Collaborative Technologies and Organizational Learning.* London, Idea Group.

Nonaka, I. and Takeuchi, H. (1995) *The Knowledge-Creating Company: How Japanese Companies Create the Dynamics of Innovation.* New York, Oxford University Press.

Nonaka, I., Toyama, R. and Konno, N. (2001) SECI, ba and leadership: a unified model of dynamic knowledge creation. In: *Managing Industrial Knowledge: Creation, Transfer and Utilization* (eds K. Nonaka, and D.J. Teece). London, Sage, pp. 13–43.

Polanyi, M. (1962) *Personal Knowledge: Towards a Post-Critical Philosophy.* London, Routledge and Kegan Paul.

Polanyi, M. (1967) *The Tacit Dimension.* London, Routledge and Kegan Paul.

Robertson, M., Sørensen, C. and Swan, J. (2001) Survival of the leanest: intensive knowledge work and groupware adaptation. *Information Technology and People* **14** (4): 334–352.

Rogers, E.M. (1983) *Diffusion of Innovations*, 3rd edn. New York, The Free Press.

Rosenberg, A. (1994) What is the cognitive status of economic theory. In: *New Directions in Economic Methodology* (ed R.E. Backhouse). London, Routledge, pp. 216–235.

Schneider, B. and Bowen, D. (1995) *Winning the Service Game.* Boston, Harvard Business School Press.

Sexton, M.G. and Barrett, P.S. (2003a) A literature synthesis of innovation in small construction firms: insights, ambiguities and questions. *Construction Management and Economics: Special Issue on Innovation in Construction* **21** (September): 613–622.

Sexton, M.G. and Barrett, P.S. (2003b) Appropriate innovation in small construction firms. *Construction Management and Economics: Special Issue on Innovation in Construction* **21** (September): 623–633.

Simon, H. (1957) *Administrative Behaviour.* New York, Macmillan Press.

Sverlinger, P.M. (2000) Managing knowledge in professional service organisation: technical consultants serving the construction industry. *PhD Thesis*, Department of Service Management, Chalmers University of Technology, Göteborg, Sweden.

Tapscott, D., Ticoll, D. and Lowy, A. (2000) *Digital Capital: Harnessing the Power of Business Webs.* Boston, Harvard Business School Press.

Wilson, T.L. (1997) Segment profitability of the US business services sector: some reflections on theory and practice. *International Journal of Service Industry Management* **8** (5): 398–413.

13 Client-oriented contractor innovation

Jan Bröchner

13.1. Introduction

In one of their pioneering studies of construction innovation, Nam and Tatum (1989) stated that 'typically, owners do not demand innovations in their facilities' but nevertheless went on to claim that the owner's demands are the key initiator of the innovation process for a construction product. Their subsequent study of ten innovative US projects from the late 1980s led them to identify another pattern: designers and contractors with advanced technologies sometimes initiated the innovative process, influenced the owners to realise their old or hidden demands, and further shaped the owner's demands (Nam and Tatum, 1992). However, a study of innovations reported in Dutch trade magazines during the twentieth century indicates that contractors had provided little more than one innovation in ten (Pries and Dorée, 2005). This figure might be explained because of a trade magazine bias towards technology innovation. And Bossink (2004) in his investigation of drivers of innovation in construction networks found that 'market pull did not have an innovation driving effect'. But was this just because he studied ten construction projects with innovations in the particular field of sustainability?

Gann and Salter (2000) have called for a better understanding of how project-based, service-enhanced firms could link project and business processes. This requires a set of concepts that goes beyond technological innovation, in particular to handle knowledge flows. What client focus means to small construction firms emerges from the study reported by Sexton *et al.* (2006): the important determinant of unsuccessful exploitation of innovations was identified as 'lack of market focus'; incidentally, the successful innovations in their survey were informational and relational, with introduction of cordless power tools being the exception, and all the unsuccessful innovations concerned tangible technologies. It seems that a typology of innovations, acknowledging a range where new tangible products are just one extreme, is needed.

In the third version of the Oslo Manual, the Organisation for Economic Co-operation and Development (OECD, 2005) has redefined innovation in order to recognise the growing importance of service innovation in developed economies. Our understanding of the service sector has improved in recent years with access to better data and theoretical advances such as the introduction of characteristics-based theories for service innovation. This gives an opportunity to take a fresh look at contractor innovation. The purpose of this chapter is, therefore, to study how contractor perceptions of client benefits from innovation affect their level of innovation activity.

13.2. Why service firms innovate

Reichstein *et al.* (2005) have analysed the construction sector responses from the 2001 UK innovation survey, comparing these responses with low-tech as well as with high-tech manufacturing. Their definition of the construction sector includes contractors and also architecture, engineering consultancy and urban planning. The analysis covers innovative performance (for product innovation and process innovation), factors hampering innovation (economic, internal, other factors) and sources of knowledge for innovation (internal, market, institutional, other, specialised). They found that construction, in particular the smaller firms, resembled traditional service industries, and they concluded that service industries provide a body of knowledge about innovation that may be 'extremely useful' for improving innovative performance in construction.

What is known about the reasons that motivate service firms to engage in innovation? The 2001 UK innovation survey is closely related to the European Community Innovation Surveys, CIS. Tether (2003) analysed CIS-2, the second European Community Survey, carried out in 1997, which included ten aims and objectives of innovation. Using data from France, Germany and Ireland together with the UK and reducing the complexity in dealing with these ten variables, he was able to find three factors for service industries, and these factors can be summarised as (i) cost reduction and regulatory compliance, (ii) service enhancement and (iii) market expansion. Reichstein *et al.* (2005) concentrated on 'hampering factors', and their analysis appears to show that innovation in construction is less inhibited by 'lack of information on markets' and 'lack of customer responsiveness to new goods or services' than what they had found in any other sectors of the economy. Responsiveness to clients is thus likely to be found as a significant mechanism among construction contractors.

The Madrid Survey on Service Innovation, carried out in 2002 and 2003, covered ten types of services including engineering and architectural services, but unfortunately not construction contracting (Gago and Rubalcaba, 2007). Among the service quality impacts of innovation, the authors distinguish between five types: flexibility in adjusting to customer needs, delivery speed, temporal availability, service user friendliness, and reliability. Their results underline the importance of ICT as a source of innovation in services, in particular because ICT is a general purpose technology that has a significant impact on the capacity to adapt to changing customer needs as well as on service user friendliness. In a related paper, the same authors (Rubalcaba and Gago, 2006) exploited data on perceived innovation impacts from CIS-3, the third Community Innovation Survey. In services, as opposed to the manufacturing industry, they showed that clients are much more important for innovations with a quality impact, while the effect is weaker for the production flexibility impact dimension and for the labour cost impact.

The Mannheim Innovation Panel data have been studied for knowledge-intensive business services (KIBS) in Germany by Hipp and Grupp (2005). Their statistical analysis reduced the effects of service innovation to four factors:

(1) Improvement of the quality of the service product.
(2) Compliance with environmental standards and safety requirements.
(3) Company internal improvements.
(4) Improvement of customer performance or productivity.

Here, they found a significant difference between KIBS and other service companies for environmental and safety requirements, which the KIBS comply with to a greater extent. Earlier, the 1997 data set for the Mannheim Innovation Panel had been analysed for the construction sector by Cleff and Rudolph-Cleff (2001). Among innovative construction firms, they discovered that clients were the most important co-operation partner, but co-operation with universities and research institutes was much lower than with manufacturing firms, whereas construction firms were clearly more likely to co-operate with their competitors. Also, competitors were the most significant source of information for innovating construction firms, followed by suppliers and clients.

Other studies of construction contractors add further insights into the structure of drivers of innovation. Mitropoulos and Tatum (2000) found that client specification of project control technologies such as scheduling tools, cost control systems and e-mail contributed as an external driver for contractor adaptation of electronic data interchange and computer-aided design. Bossink (2004), who interviewed 66 construction experts in the Netherlands, pointed out four types of innovation drivers: environmental pressure, technological capability, knowledge exchange and boundary spanning.

13.3. Types of innovation

If it is important to distinguish between types of innovation and types of service characteristics, the theoretical framework proposed by Gallouj and colleagues (Gallouj, 2002) allows mapping of the innovation dynamics of construction contractors. In this framework, which has been applied to hospitals (Djellal and Gallouj, 2005) and other contexts where there is a mix of tangible and intangible production, four variables or rather vectors are distinguished: (1) constituent services, (2) service mediums, (3) service characteristics (utilities) and (4) competences. For the constituent services, competences are mobilised according to service mediums to change the bundle of service characteristics that are related to customer utility functions. Innovation can also take the form of adding or removing constituent services. In the present context, the constituent services are interpreted as the principal activities of construction contractors.

Moreover, four service trajectories related to service mediums are identified within this framework: Material, Information, Knowledge and Relational. The first, Material, stands for technology and is a logistical and material transformation trajectory, which is what earlier surveys of construction innovation often have been restricted to. The second, Information, is also a logistical trajectory but even more so a data processing trajectory. The third, Knowledge, trajectory concerns methodological or intellectual methods; it can be interpreted as relating to the introduction of new routines, e.g. quality assurance in the construction firm. The fourth and last trajectory, Relational, is a pure services trajectory that arises from the direct mobilization of a competence to provide a particular service function or characteristic. The development of partnering relations to construction clients is suggested as an example of the relational trajectory.

The large Canadian and Australian construction innovation surveys (Anderson and Schaan, 2001; Seaden *et al.*, 2003; Manley, 2005) with their lists of advanced technologies and business practices allow interpretation in terms of the four service trajectories. Thus interpreting the lists given by Manley and McFallan (2006), the Material trajectory would correspond to their 'Materials' and parts of both 'Plant and equipment' and 'Systems', while Information covers 'Computerised practices' and those elements which do not fall under Material among 'Plant and equipment' and 'Systems'. 'Design' is divided between the Information trajectory and the Knowledge trajectory, where most of 'Organisational practices' are found. Finally, the Relational trajectory is where 'Contracts' belong and what remains of 'Organisational practices'.

13.4. Client satisfaction

Vitruvius might be said to have settled the three types of product quality for buildings – firmness, commodity and delight – but there is less agreement on how to structure the process qualities that cause satisfaction among construction clients. The importance of process service aspects is brought out by a study of Florida home buyers (Torbica and Stroh, 2001). Within the services innovation framework that has just been presented, services are described partly by characteristics that impact on customer utility levels. The general service quality dimension approach (Parasuraman *et al.*, 1988) which – with some influence from the literature on what constitutes project success – has led to several construction sector applications with an empirical base: for US contractors (Ahmed and Kangari, 1995), Jordanese contractors (Al-Momani, 2000), a Swedish housing refurbishment contractor (Holm, 2000), for US electrical contractors (Maloney, 2002) and also for Finnish contractors (Kärnä, 2004).

The UK key performance indicators (KPIs) for construction give useful pointers (e.g., client satisfaction – product, defects at handover, construction time; predictability – time, client satisfaction – service) (KPI Working Group, 2001). Using KPIs for monitoring does raise issues of measurability that are even more pronounced for procurement criteria: the rise of multiple, non-price criteria for the award of public contracts is another source of knowledge about client priorities, including environmental concerns (Waara and Bröchner, 2005, 2006). In Table 13.1, the eight service characteristics chosen for the survey are presented together with examples from earlier studies. These eight characteristics are to be understood as perceived external impacts, and a categorisation of six internal impacts within the contractor's organisation was also prepared.

13.5. The Swedish survey

The target population of the survey about to be described is the 50 construction contractors that in 2004 had the highest turnover in their Swedish construction activities. With support from the Development Fund of the Swedish Construction Industry, a survey questionnaire was developed with a first part containing firm-level questions and a second part with questions for each principal activity the contractor is engaged in. Most questions were to be answered yes/no, and questions requiring answers on degree scales or the ability to formulate individual answers were avoided because of adverse experiences from earlier surveys of construction (Anderson, 2005). Questionnaires were distributed in 2006 and the time period covered was 3 years in retrospect, 2003–2005. Table 13.2 indicates the size distribution and response rates for the survey; the average response rate is high, 88%.

Of all 44 firms, 30 reported that they had conducted R&D during the period, 26 that they had made an innovation new to the firm, and 11 that they had made at least one innovation that they perceived as being new to the country. Only one firm indicated that

Table 13.1 Service characteristics of construction projects

Service characteristics (source of customer satisfaction) in the Swedish survey	Examples of similar indicators
Higher technical quality	Firmness (Vitruvius); quality (Ahmed and Kangari, 1995); quality (Holm, 2000); client satisfaction – product (KPI Working Group, 2001)
Higher technical quality precision	Quality assurance and handover (Kärnä, 2004); defects at handover (KPI Working Group, 2001)
Shorter time	Construction time (KPI Working Group, 2001)
Higher time precision	Time (Ahmed and Kangari, 1995); predictability – time (KPI Working Group, 2001)
Higher esthetical quality	Delight (Vitruvius)
Reduced local negative environmental effects	Environment and safety (Kärnä, 2004)
Reduced global negative environmental effects	Environment (Waara and Bröchner, 2005)
Higher ability to communicate with client	Communication, response to complaints (Ahmed and Kangari, 1995); project flexible (Al-Momani, 2000); contractor/customer relationship (Maloney, 2002); co-operation (Kärnä, 2004)

they had made an innovation (new to the firm) without engaging in R&D. As shown in Table 13.3, the 44 firms submitted information on a total of 173 activities belonging to 17 types of activities, obviously corresponding to a great diversity of customers.

Firms were asked to report on three levels: R&D efforts, as defined in the Frascati Manual (OECD, 2002), innovations new to the firm, and innovations that they perceived to be new to the country, relying on the Oslo Manual definition of innovation

Table 13.2 Swedish contractors in the 2006 survey (2004 data)

Number of employees	Total number in population	Responding firms	Average (MSEK) turnover/employee for responding firms
10–49	2	2	15.3
50–249	36	30	2.9
250–499	5	5	1.5
500–999	1	1	2.7
1000–4999	3	3	2.2
5000 and above	3	3	2.0
All firms	50	44	3.2

Table 13.3   Contractor activities (firms, $N = 44$)

Activity	NACE code Rev. 2	Number of firms 2005
Quarrying of stone, sand and clay	08.1	5
Manufacture of construction materials and articles	16 + 23	5
Water supply, sewerage, waste management and remediation activities	36 – 39	2
Development of building projects	41.1	21
Construction of residential buildings	41.21	30
Construction of non-residential buildings	41.22	32
Construction of roads and railways	42.1	16
Construction of utility projects	42.2	11
Construction of other civil engineering projects	42.9	15
Demolition and site preparation	43.1	11
Electrical, plumbing and other construction installation activities	43.2	7
Building completion and finishing	43.3	4
Other specialised construction activities	43.9	8
Real estate activities	68	8
Architecture and engineering activities; technical testing and analysis	71	9
Renting and leasing of machinery, equipment and tangible goods	77.3	9
Services to buildings and landscape activities; office administrative and support activities	81 – 82	8

(OECD, 2005). Given the peculiarities of construction markets, the expression 'new to the country' was preferred to 'new to the market' in the questionnaire.

The questionnaire used the Swedish term *kund* (customer) when referring to construction clients, although *byggherre* (approximately = client) was added parenthetically as an explanation. Here, the Swedish terminology reflects the continental tradition of seeing the client as the lord or master of a project: *Bauherr* in German, *maître de l'ouvrage* in French. Collaborative competences were categorised according to eight types of partners, including clients, in R&D and innovation. Contractors were asked to indicate which sources of customer satisfaction (= the eight service characteristics already identified) they thought that their R&D or innovation activities had influenced.

Table 13.4   R&D, innovation and activities by size of firm

Number of employees	Firms reporting R&D	Firms reporting innovation 'new to the firm'	Firms reporting innovation 'new to the country'	Average number of activities
10–249	20	18	6	3.9
250–499	3	2	0	4.0
500 and above	7	6	5	8.1
All firms	30	26	11	4.6

13.6. Results

In this section, the main results are presented. Reducing the firm size ranges in Table 13.2 to only three, basic data on R&D, innovations and activities (as in Table 13.3) are given in Table 13.4.

On average, each respondent firm was diversified into 4.6 types of principal activities according to Tables 13.3 and 13.4. Innovations in the form of activity entry and exit movements appear to have been insignificant during the 3 year period investigated, and there is no readily ascertained pattern for entry and exit.

For the 133 areas, clients were the most (68%) frequent type of collaborator, followed by consultants (60%), other type of contractor (43%), IT suppliers (41%), same type of contractor (35%), equipment suppliers (32%), universities or research institutions (31%) and suppliers of materials (29%). When compared to first-time clients, recurrent clients were perceived by most respondents to be more willing to initiate or to accept innovations. Continuity is also reflected in responses to the question whether the firm's most important innovation had been applied in more than one project: all but two of the firms who had reported at least one innovation agreed that this was the case. Another question was whether public sector clients were important for R&D and innovations within the firm: two thirds of firms with more than 1000 employees thought so, but this opinion was shared by less than half of the smaller firms, and 38% of these thought that clients in the public sector had a negative influence.

The figures in Table 13.4 can be analysed according to the service trajectories (Material, Information, Knowledge or Relational) followed by contractors. In the questionnaire, it was possible to indicate a single trajectory or a combination of trajectories. Results are found in Table 13.5. While the knowledge trajectory dominates R&D efforts, probably to be seen as almost exclusively being development, the spread across the four trajectories is more even for innovations new to the firm, and with the stricter definition of new to the country, it is the material trajectory that reaches the top position.

Thus Table 13.5 shows how strong the relative decline of the knowledge trajectory is when successively stricter definitions of innovation are imposed. However, it can also be seen that the material trajectory fails to dominate the third group of activities.

For the analysis, an innovation intensity variable is defined with the following values: Innovation new to the country (= 3); Innovation new to the firm (= 2), R&D (= 1), no R&D or innovation (= 0); the trajectory is categorised as Material, Information, Knowledge or Relational. Since it is unlikely that this intensity scale is linear, linear regression would be inappropriate for determining what affects innovation intensity.

Table 13.5   Level of R&D and innovation among surveyed contractors by trajectory (number of firms, $N = 44$)

	Trajectory			
Level of R&D and innovation	**Material**	**Information**	**Knowledge**	**Relational**
Research and development	19	19	28	17
Innovation new to the firm	15	13	16	10
Innovation new to the country	8	4	6	4

In Table 13.6, ordinal regression with a logit link function (Liu and Agresti, 2005) has been applied with innovation intensities as the four dependent variables.

Data are taken from the 133 principal activity areas with non-zero innovation intensity in the 44 firms. The goodness-of-fit estimates (Nagelkerke's pseudo R^2) indicate that the predictive power of the model is especially strong for relational innovation intensity. In Table 13.6, it might surprise that 'higher ability to communicate with client' is significantly associated with the material trajectory and not with the informational, which should be the natural candidate. However, closer analysis of contractor collaboration patterns on each of the levels of innovation intensity reveals a more complex picture: having IT suppliers as partners is usual for development work along the information trajectory, but when it comes to innovations new to the country, suppliers of materials and components are more important for the information trajectory – even more so than for the material trajectory.

Turning to the relative importance of the eight service characteristics listed in Table 13.6, there are conspicuous differences in the assessment of esthetical quality, where traditional construction activities are less influenced, by environmental considerations. For all three groups of activities, higher ability to communicate with client is given a high rank, but less often so for traditional construction. For the third group of activities, 'shorter time' is a relatively less important effect, but this is only what should be expected for activities that typically provide services for a fixed period that might be subject to prolongation for at least another fixed period.

The customer satisfaction effects (as in Table 13.6) can be studied in their relation to internal effects (Table 13.7) in the contractor firm, as perceived by respondents. Strong correlations (>0.5) are found between the external impact variable 'higher technical quality' and the internal impact 'employee health and safety'. There is a broader relationship between 'higher technical quality precision' and internal variables except for 'image' and 'production efficiency'. 'Shorter time' has its strongest correlation going upstream in the supply chain ('better relation to subcontractors or suppliers'), and this link is even stronger for 'higher time precision'. Reduction of local negative environmental effects is correlated with the two relationship impacts, both the employee relations and the supply chain relations. But the global environmental effects are clearly linked to the company image impact.

13.7. Conclusion

The survey analysis indicates that Swedish construction contractors are aware of a variety of client preferences when they spend resources on R&D or innovation, and

Table 13.6 Ordinal regression estimates of innovation intensity in business areas performing R&D or innovating ($N = 133$) (standard errors in parentheses)

Variable	Trajectory			
	Material	**Information**	**Knowledge**	**Relational**
Threshold				
No R&D or innovation	1.993***	1.227**	−0.197	2.635***
	(0.418)	(0.368)	(0.365)	(0.489)
R&D	2.819***	2.480***	2.314***	4.226***
	(0.458)	(0.421)	(0.401)	(0.590)
Firm innovation	3.775***	3.882***	3.987***	5.385***
	(0.509)	(0.513)	(0.512)	(0.668)
Location				
Higher technical quality	2.195***	0.303	0.839*	0.897
	(0.462)	(0.425)	(0.408)	(0.496)
Higher technical quality precision	−1.216*	−0.619	0.101	0.811
	(0.503)	(0.468)	(0.429)	(0.512)
Shorter time	0.949	1.914***	0.399	1.990***
	(0.521)	(0.499)	(0.480)	(0.535)
Higher time precision	0.784	0.825	1.039*	0.394
	(0.499)	(0.482)	(0.479)	(0.531)
Higher esthetical quality	0.512	0.904	0.398	1.454*
	(0.543)	(0.561)	(0.564)	(0.595)
Reduced local negative environmental effects	−0.719	0.577	0.949*	1.574**
	(0.504)	(0.453)	(0.436)	(0.462)
Reduced global negative environmental effects	0.801*	1.138**	1.843***	0.045
	(0.403)	(0.383)	(0.393)	(0.009)
Higher ability to communicate with client	1.093*	−0.195	−0.436	−0.077
	(0.429)	(0.425)	(0.400)	(0.491)
−2 log likelihood	209.787	215.599	238.638	180.184
X^2	53.346	72.866	74.466	108.553
Nagelkerke pseudo R^2	0.361	0.454	0.462	0.609

*** $p < 0.001$; ** $p < 0.01$; * $p < 0.05$

that perceived client preferences have different profiles according to type of innovative activity. In general, the analysis presents evidence that the new, broader OECD innovation definition that abandons an exclusive focus on product and process technology reduces the gap between the construction industry and other industries, in particular within the service sector. Construction in advanced economies thus appears to join the new mainstream of project dependent service industries, and its pattern of innovation

Table 13.7 Correlations between external and internal impacts (133 activity areas)

External impact	Internal impact					
	More efficient production process	More stable capacity use	Better occupational health and safety for employees	Better leadership and relations between employees	Better relation to subcontractors or suppliers	Better image of the firm
Higher technical quality	0.385	0.310	0.607	0.420	0.295	0.145
Higher technical quality precision	0.366	0.491	0.512	0.567	0.504	0.024
Shorter time	0.475	0.469	0.413	0.435	0.528	0.172
Higher time precision	0.385	0.479	0.504	0.380	0.683	0.257
Higher esthetical quality	0.138	0.277	0.262	0.115	0.358	0.220
Reduced local negative environmental effects	0.350	0.424	0.464	0.524	0.536	0.277
Reduced global negative environmental effects	0.031	0.339	0.143	0.333	0.326	0.561
Higher ability to communicate with client	0.038	0.350	0.160	0.214	0.352	0.243

can be interpreted accordingly. However, the geographical dispersal of construction sites, the large flows of materials and dependence on public procurement practices in many markets imply that incentives for construction innovation will differ somewhat.

Another obvious question raised by the survey is how contractor perceptions agree with actual preferences among Swedish construction clients. The fact that frequent client–contractor collaboration in innovatory work is indicated by respondents might not be sufficient to ensure that perceived and actual preferences coincide. Caution is needed even more when assessing the global applicability of results obtained from this survey. Miozzo and Dewick (2004, p. 58), who interviewed the largest three or four contractors in five European countries, emphasise how the major Swedish contractors have 'very strong' collaborations with clients. In cultures with greater gaps between contractors and clients, there might be significant divergence between what contractors believe is vital to clients and how clients themselves experience process qualities in construction projects.

Although the general understanding of how service industry innovators perceive the impact of their innovations is developing, there is not yet any established framework for systematising client satisfaction variables. Impact patterns might be strongly determined by technologies and market features so that construction – or any specific type of construction – needs its own sets of factors; a good topic for research.

References

Ahmed, S.M. and Kangari, R. (1995) Analysis of client-satisfaction factors in construction industry. *Journal of Management in Engineering* **11** (2): 36–44.

Al-Momani, A.H. (2000) Examining service quality within construction processes. *Technovation* **20**: 643–651.

Anderson, F. (2005) Measuring innovation in construction. In: *Building Tomorrow: Innovation in Construction and Engineering* (eds A. Manseau and R. Shields). Aldershot, Ashgate, pp. 57–80.

Anderson, F. and Schaan, S. (2001) *Innovation, Advanced Technologies And Practices in the Construction and Related Industries*. Statistics Canada Working Paper, January.

Bossink, B.A.G. (2004) Managing drivers of innovation in construction networks. *Journal of Construction Engineering and Management* **130**: 337–345.

Cleff, T. and Rudolph-Cleff, A. (2001) Innovation and innovation policy in the German construction sector. In: *Innovation in Construction: An International Review of Public Policies* (eds A. Manseau and G. Seaden). London, Spon, pp. 201–234.

Djellal, F. and Gallouj, F. (2005) Mapping innovation dynamics in hospitals. *Research Policy* **34**: 817–835.

Gago, D. and Rubalcaba, L. (2007) Innovation and ICT in service firms: towards a multidimensional approach for impact assessment. *Journal of Evolutionary Economics* **17**: 25–44.

Gallouj, F. (2002) *Innovation in the Service Economy: The New Wealth of Nations*. Cheltenham, Edward Elgar.

Gann, D.M. and Salter A.J. (2000) Innovation in project-based, service-enhanced firms: the construction of complex products and systems. *Research Policy* **29**: 955–972.

Hipp, C. and Grupp, H. (2005) Innovation in the service sector: the demand for service-specific innovation measurement concepts and typologies. *Research Policy* **34**: 517–535.

Holm, M.G. (2000) Service quality and product quality in housing refurbishment. *International Journal of Quality and Reliability Management* **17**: 527–540.

Kärnä, S. (2004) Analysing customer satisfaction and quality in construction: the case of public and private customers. *Nordic Journal of Surveying and Real Estate Research, Special Series* **2**: 67–80.

KPI Working Group (2001) *KPI Report for the Minister for Construction.* UK Department of the Environment, Transport and the Regions, January.

Liu, I. and Agresti, A. (2005) The analysis of ordered categorical data: an overview and survey of recent developments. *Test* **14**: 1–73.

Maloney, W.F. (2002) Construction product/service and customer satisfaction. *Journal of Construction Engineering and Management* **128**: 522–529.

Manley, K. (2005) *BRITE Innovation Survey.* Cooperative Research Centre (CRC) for Construction Innovation, Brisbane, Australia.

Manley, K. and McFallan, S. (2006) Exploring the drivers of firm-level innovation in the construction industry. *Construction Management and Economics* **24**: 911–930.

Miozzo, M. and Dewick, P. (2004) *Innovation in Construction: A European Analysis.* Cheltenham, Edward Elgar.

Mitropoulos, P. and Tatum, C.B. (2000) Forces driving adoption of new information technologies. *Journal of Construction Engineering and Management* **126**: 340–348.

Nam, C.H. and Tatum, C.B. (1989) Toward understanding of product innovation process in construction. *Journal of Construction Engineering and Management* **115**: 517–534.

Nam, C.H. and Tatum, C.B. (1992) Strategies for technology push: lessons from construction innovations. *Journal of Construction Engineering and Management* **118**: 507–524.

OECD (2002) *Frascati Manual: Proposed Standard Practice for Surveys on Research and Experimental Development*, 6th edn. Paris, OECD.

OECD (2005) *Oslo Manual: Guidelines for Collecting and Interpreting Innovation Data*, 3rd edn. Paris, OECD.

Parasuraman, A., Zeithaml, V.A. and Berry, L.L. (1988) SERVQUAL: a multiple-item scale for measuring consumer perceptions of service quality. *Journal of Retailing* **64**: 12–40.

Pries, F. and Dorée, A. (2005) A century of innovations in the Dutch construction industry. *Construction Management and Economics* **23**: 561–564.

Reichstein, T., Salter, A.J. and Gann, D.M. (2005) Last among equals: a comparison of innovation in construction, services and manufacturing in the UK. *Construction Management and Economics* **23**: 631–644.

Rubalcaba, L. and Gago, D. (2006) Economic impact of service innovation: analytical framework and evidence in Europe. Paper presented at the *IoIR/ASEAT Conference on Innovation in Services*, 15–17 June, Manchester.

Seaden, G., Guolla, M., Doutriaux, J. and Nash, J. (2003) Strategic decisions and innovation in construction firms. *Construction Management and Economics* **21**: 603–612.

Sexton, M., Barrett, P. and Aouad, G. (2006) Motivating small construction companies to adopt new technology. *Building Research and Information* **34**: 11–22.

Tether, B.S. (2003) The sources and aims of innovation in services: variety between and within sectors. *Economics of Innovation and New Technology* **12**: 481–505.

Torbica, Ž.M. and Stroh, R.C. (2001) Customer satisfaction in home building. *Journal of Construction Engineering and Management* **127**: 82–86.

Waara, F. and Bröchner, J. (2005) Multicriteria contractor selection in practice. *Proceedings of the CIB W92 International Symposium on Procurement Systems*, 8–10 February, Las Vegas, pp. 167–172.

Waara, F. and Bröchner, J. (2006) Price and nonprice criteria for contractor selection. *Journal of Construction Engineering and Management* **132**: 797–804.

14 Driving innovation in construction: a conceptual model of client leadership behaviour

Mohammed F. Dulaimi

14.1. Introduction

There is no doubt that the involvement of the client/customer in the development of a new product or service is essential for its success. It is also evident that the current interest of business organisations to embrace a more customer-oriented culture is bringing the attention of the supply chain and individuals to the importance of the client input. The motivation to become customer-oriented is the belief that such organisations are able to reduce the risk of failure of new products and services through the effective definition of customer needs and expectations. This is no different to what is happening in the construction industry.

The pressure to accelerate change in the construction industry requires a more radical change in order to embrace innovations that can help the industry to achieve the desired step improvements. Radical innovation in products and services may bring significant growth but also significant risk to the instigator of innovation. Such risk could discourage the other parties in the construction project without strong leadership from the client that would inspire and sustain successful innovations. This chapter will propose a framework for clients' leadership to promote innovation in construction projects. This framework can be used to inform clients and their advisors of the opportunities for clients to take a more pro-active role throughout the life cycle of the project.

14.2. The motivation for a new approach

The rapid pace of increased competition in the last two decades has motivated business organisations to explore new philosophies that would give them the edge over their competitors. Quality assurance, total quality management, re-engineering, value management and lean thinking have crept into the business development agenda of many organisations. However, such new ideas have only allowed firms to achieve, in most cases, incremental improvements to existing practices that have enabled such firms to compete in existing markets. In a very crowded market corporate success cannot be sustained without a sustainable innovation strategy that continuously creates and exploits new markets rather than competing in existing markets. Such innovations would create a leap of value for customers that creates not only a significant gap that is difficult for competitors to bridge in the immediate future but would also create new rules that define success in the new market. To remain successful such firms need to rapidly exploit newly discovered markets before competitors catch up.

Kim and Mauborgne (2005) described these markets as 'Blue Oceans' where competition becomes irrelevant. The results of their research of successful organisations over the last 100 years indicate that many of the well-known successful organisations were not perpetually successful. These organisations were successful because they were able to implement 'strategic moves' that have delivered projects, products and services that have created a new market – Blue Ocean – where there is no competition. The fortunes of these companies changed when they stopped making such products. For example, the automobile industry at the first half of the 20th century saw the rise of the Ford Corporation on the back of their revolutionary T Model that provided the first affordable car. However, Ford's fortunes changed as General Motors came up with a new product, a strategic move, which offered not only affordability but different styles as well (Kim and Mauborgne, 2005). In construction, the pioneering work of Nakheel in the United Arab Emirates of building 'over water' has created a success story that many developers in the region were quick to follow. Nakheel were very quick to exploit this new concept by pushing further to create a significant gap between itself and any likely competitor. Nakheel launched the 'Palm', then the 'World' and recently the 'Arabian Canal' projects generating massive mixed-use developments. In all the three developments Nakheel managed to identify a Blue Ocean where none of its competitors have been.

Such strategic moves create products that 'teach' customers new needs and thus create new markets for the new product. Such companies were able to do so by challenging the industry logic and hence exploring areas where the industry never ventured. Such organisations succeeded by creating the environment and providing the leadership that motivated their supply chain to bring these innovations into existence.

In order to realise such innovations such organisations had to overcome two major hurdles. The first, is 'defining' this new product or service and secondly assembling a supply chain that can be motivated to work in an integrated manner to realise this innovation. The driving hypothesis of this research is that client leadership is most effective in defining and realising successful innovations. The client engagement of a highly motivated supply chain will create greater opportunities for discovering new products and services. Here the client plays a pro-active role in promoting and affecting a customer-oriented culture. The client should also intervene to sustain the motivation of the supply chain towards achieving the desired targets.

Hence, this chapter will propose an outline of how clients can drive innovation in construction projects by providing the leadership that inspires and sustains successful innovations. This chapter will help build the theoretical base by examining the concepts of innovation, customer orientation and the supply chain integration.

14.3. Promoting greater innovation in construction

The construction industry, worldwide, has been under pressure to emulate the major successes achieved elsewhere in the economy. In countries such as Singapore, Hong

Kong and the United Arab Emirates where the creation of a knowledge-based economy is seen to be the blue print for the country's future the construction industry finds itself under pressure to address the performance gap between itself and the rest of the economy. This gap, in many cases, is difficult to bridge without radical change. Almost all recent reviews of construction industry performance have identified the need for the industry to change and embrace a culture that promotes innovation and continual improvement. It is common to find reports and papers that criticise the construction industry for not focusing on the customer, being fragmented and not investing enough in research and development (R&D) (Egan, 1998; Construction 21, 1999).

However, most of the debate on the organisation of the construction development process tends to give the client/customer a more passive role especially beyond the concept development phase of the project. Leadership studies in the construction industry focused on the leadership provided by the different professionals but not the client. It is important to develop a deeper understanding of the client role in driving successful innovation. Previous studies on construction innovation concentrated on the barriers and enablers of innovation (Sze *et al.*, 2005) and the role of key professionals such as project managers in championing innovation (Park *et al.*, 2004)

14.4. The innovation dimension

Research by Dulaimi *et al.* (1996) has shown that the development process in construction from concept to detailed design and construction is similar to the process of progressing from R&D to successful innovations in manufacturing scenarios. It has been argued that the structure and organisation of activities in an R&D environment are comparable to existing project organisational models and procurement frameworks deployed in construction projects. Both processes, essentially, aim to find a solution to a problem or a need. They also share similar aspirations to identify solutions to a need or a problem.

One of the main issues facing new product development is how to reintegrate what has been divided as organisations and their products become more complex (Liker *et al.*, 1999). The bringing together of the design and manufacturing/implementation processes used to occur inside individual minds. Failure to integrate these processes will undermine innovation.

The realisation of new ideas results from the collaboration of two very diverse operations: R&D and manufacturing. These two operations occur in two very different environments. Those involved in the R&D phase work in an environment that fosters creativity and innovation. This environment also allows greater tolerance of uncertainty and risk. Greater effort is made by management to break down barriers, such as concerns for 'this is not how we do things here'. On the other hand, the manufacturing/production phase environment is considerably constrained by details and the need to meet deadlines and budgets. Precision planning and programming activities dominate operations by technicians and professionals in this phase.

The above description suggests that the two phases are independent with different management organisations, cultures and styles. The successful management of the development and production of new products would require integration of the two phases with management appreciating and synergising the different cultures and styles. In managing innovation, the organisation will need to link the two phases without stifling innovation and at the same time not losing sight of the needs of the future product. The new idea will need to cross successfully from R&D to manufacturing. Assigning each phase to a separate organisation, with different management priorities and objectives, will create several problems in the effective transfer of the new idea and would increase the risk of failure of this innovation. Being new and innovative this idea may be viewed by the implementer/adopter of the new idea, rather than originator, as strange and risky and may, in fact, increase the risk of failure. To rely on contractual agreements to limit their risk both the originator and implementer may face the problem of not being able to identify all potential risk aspects of a new, non-tested, idea. Alternatively, both organisations need to support each other by making technical, commercial or managerial information and advice readily available to enhance the chances of success. Hence, the effective management of the creation and transfer of knowledge within the two 'phases' are central to the success of new products or technology.

In failing to integrate, effectively, design and implementation the performance of the project will suffer. This cost can be associated not only with failing to develop the best design of the product that can be efficiently manufactured, but also with the inability to market and deliver the new product to the customer as fast as possible, at the right time, and before competitors (Susman and Dean, 1992).

The design and construction of a building or structure require the coming together of different groups and organisations. One of the main challenges to success will be whether this network of organisations will facilitate or hinder the development of more innovative solutions. There will be several paradigms competing to dominate and influence the structure, attitude and practices deployed on a particular project (Dulaimi and Kumaraswamy, 2000). The expectation of significant benefits to the different members of the supply chain from the proposed innovation should motivate the parties to exert an effort to bridge the gaps between themselves. This chapter argues that the client has/can have the power to influence a positive integration and affect synergy to a fragmented process.

14.5. Customer orientation dimension

Customer orientation has been a prominent issue in business and management literature for long time. Drucker (1954) suggested that what a business thinks it produces is not of first importance – especially not to the future of the business and to its success. Drucker affirms that what the customer thinks he is buying, what he considers value is decisive. The growth of customer orientation in management and operation continues across almost every industry, manifested in part in the growth of relationship marketing as a strategic policy for many organisations (Christopher and McDonald, 1995). A customer-driven business develops a comprehensive understanding of its customers' business and how customers in the immediate and downstream markets perceive value.

Narver and Slater (1990, p. 21) stated that 'market orientation is the organisation culture that most effectively and efficiently creates the necessary behaviour for the creation of superior value for buyers and, thus, continuous performance for the business'. Dulaimi (2005) has adapted a model to describe the customer-oriented approaches in construction. The model has three main components: intelligence generation, dissemination of intelligence and organisation wide responsiveness. Dulaimi (2005) argued that this proposed model should be able to aid the project team to develop the concept and details of the services and product that would effectively satisfy customer needs and expectations. This model clearly gives the leadership in influencing the product definition and process of development to professional advisers and consultants. The major threats to any initiative to create an effective customer-oriented culture are the fragmented nature of the knowledge created; the 'localised' nature of the response, and the lack of senior management support of change.

14.6. The motivation dimension

As construction is a project-based activity, the organisational context of innovation is characterised by a specific temporary alliance of independent organisations with different interests and capacities (Slaughter, 1998; Gann and Salter, 2000). Close resource and information exchange between the organisations involved in a construction project is one of the prerequisites of successful innovation. The project procurement path lays down broad rules and creates the working environment that would facilitate such a resource and information exchange, and helps adjust the project context. Different procurement paths possess relatively different sets of rules and, therefore, result in different working environments.

Irrespective of the adopted procurement path in a project there are some fundamental underlying mechanisms that affect the level of support and collaboration. In an attempt to implement a new idea on a construction project, one party could be motivated and committed towards the idea with the expectation of attaining certain explicit or implicit targets or goals. During and after implementation of innovation, satisfactory or unsatisfactory progress and end results of performance would motivate or demotivate the different parties and individuals to exert further efforts for this innovation.

Dulaimi *et al.*'s (2002) study of the above issues raised the question of how can the alignment of motivation in the supply chain on construction projects be achieved. Even if the management of the different organisations in the supply chain are committed to the new idea, this commitment will be challenged to gain operational commitment in their respective organisations. It takes time and resources to prepare the operational body of each organisation to implement the new idea. Integration and coordination efforts within the supply chain will be challenged to harness consistent effort from the chain. Inconsistent effort would produce unfavourable results that may not be conducive to enhance further commitment and further effort. Even if the results are favourable there is less chance that the results would stimulate further implementation of the same innovation.

14.7. The leadership dimension

Nearly all studies on leadership in the construction industry have focused on the role of the different professionals, mainly the project/site manager in providing leadership to the construction project team. Bresnen *et al.*'s (1987) research at Loughborough University, UK, explored the socio-psychological aspects of management of projects focusing on the motivation and leadership aspects of site management. The work of Dulaimi (1991) and Dulaimi and Langford (1999) developed behavioural measures to identify the impact of personal and situational variables on project managers' behaviour. Fraser (2000) studied the impact of personal factors on site managers' effectiveness. Briscoe *et al.* (2004) provided a rare opportunity, using case studies, to explain how client leadership can influence supply chain integration. In addition to studies in non-construction set-ups, researches in construction have advocated the involvement of the client to be key to the success of innovation (Dulaimi and Kumaraswamy, 2000).

14.7.1. The client as 'controller' of the project

The client is the main party in the project concerned with ensuring that the behaviour and activities of all project stakeholders contribute to meeting his/her goals and objectives. Therefore, the client needs to manage the project stakeholders to elicit the commitment and contribution that would deliver the desired results through his/her exercise of controls, formal and informal. Hence, it is important to deploy a portfolio of controls to enable the client to motivate all participants and facilitate their working towards the set objectives. In addition, it should be recognised that there will be several moderating influences to such an endeavour that relates to the client sophistication and expertise, the nature of the product or service being procured and the project and business environments. However, at this stage this research focus is on building the framework for client leadership.

'Control' is defined in behavioural terms to mean the endeavour to ensure that individuals, groups and organisations contributing to the project would act according to an agreed strategy to achieve the desired objectives. Kirsch (1997) and Kirsch *et al.* (2002) identified four modes of control – two formal and two informal. The formal modes of control are behaviour and outcome. Here, the client defines the procedure and tasks in the former and the outputs in the latter. The client, then, evaluates performance to the extent to which the different participants in the project have adhered to the set procedures and delivered a satisfactory outcome. Such modes of control will be appropriate when the appropriate behaviours and outcomes are known to the client. These controls should also be observable and can be articulated by the client (Eisenhardt, 1985). The manner in which construction procurement frameworks have evolved allows the client to exercise these controls. However, clients will need to have the knowledge, capabilities and expertise to be able to define the objectives and direct the project supply chain forwards.

The informal modes of control are clan control and self-control. Self-controls are related closely to the ability of the client to instil the necessary commitment through the alignment of motivation, at individual and group levels, towards the achievement of the desired objectives. Clan control refers to the social pressure a group can apply on individuals within the group to adhere to the group's standards and norms. The reward for conforming, in this case, is the retainment of individuals within the group and the realisation of the group's shared goals and objectives (Ouchi, 1980). The move towards more collaborative working arrangements, such as partnering, are examples of clients exercising the power of the controller of the project's key resources on the supply chain.

14.8. Discussion

The review in this chapter highlighted some of the key issues that would help enhance the level of innovation in industry. It provided directions on how the client can influence positively the culture, strategy and the process of development. This section will briefly outline an integrated behavioural framework of the client role in this context.

Research has trusted the thesis that all factors that influence job behaviour can be broadly categorised into two main categories – personal and situational (Bresnen *et al.*, 1987; Dulaimi, 1991; Dulaimi and Langford, 1999). Research over the last 50 years or so has provided a wealth of knowledge on how the personal and situational variables interact and influence job behaviour.

The personal factors refer to what the job holder, the client, brings to the project that would influence the client behaviour. The client background, experience and expertise are expected to influence the power' the client can exercise to influence the supply chain across the development stages of the project. The work of Liu *et al.* (2003) and Liu and Fang (2006) have proposed a model that would help understand how the power structure in projects evolves and its relationship to the project environment and project performance. Future research can adapt this model to focus on the client organisation and linking it to the client 'motivational hierarchy'. The development and application of such a model should provide important intelligence as to where the client intervention will produce the expected results.

The motivational drivers of the client will be influenced not only by the commercial drivers of the project and corporate strategy but also by the client business culture. The client may well be interested in bottom line solutions with lowest cost mentality or, at the other end, would be interested in identifying 'Blue Ocean' in their line of business. It is proposed that future research should focus on major corporate clients

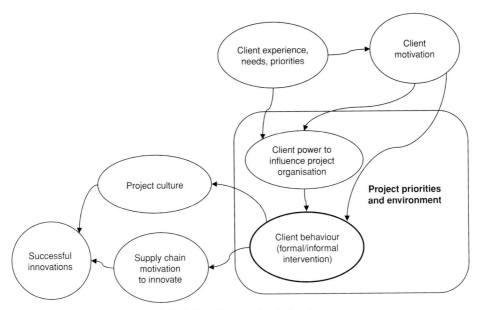

Figure 14.1   Conceptual model of client leadership behaviour.

who would have the expertise, ability and the resources to implement the research recommendations.

The other aspect of the model is the behavioural dimension. The work of Kirsch (Kirsch, 1997; Kirsch *et al.*, 2002) into control modes and client, and client representative, can be adapted to identify the behavioural dimensions that describe the leadership that clients can provide to influence a customer-oriented project culture that works in an integrated manner towards a common objective. The development of such measures should enable researchers to examine client leadership behaviour and its impact on innovation process.

In conclusion, the main thrust of this chapter has been that the client as the sponsor of the project has, can have, the power to intervene to drive innovation on construction projects (see Figure 14.1). Research is needed to develop the knowledge of what patterns of behaviour of clients are able to positively influence the culture of the project and provided the incentive for the supply chain to instigate and support the development and implementation of new ideas.

References

Briscoe, G.H., Dainty, A.R., Millett, S.J. and Neale, R.H. (2004) Client-led strategies for construction supply chain improvement. *Construction Management and Economics* **22**: 193–201.

Bresnen, M.J., Bryman, A., Beardsworth, A. and Keil, E.J. (1987) *Effective Site Management.* Occasional Paper, No. 85, Chartered Institute of Building (CIOB).

Christopher, M. and McDonald, M. (1995), *Marketing: An Introductory Text.* London, MacMillan.

Construction 21 (1999) *Reinventing Construction.* Singapore, Construction Task Force, S&P.

Drucker, P. (1954) *The Practice of Management.* New York, Harper & Row.

Dulaimi, M. (1991) Job behaviour of site managers: its determinants and assessment. *PhD Thesis,* University of Bath, UK, unpublished.

Dulaimi, M., Langford, D. and Baxendale, A. (1996) The construction project as a process of innovation. *Proceedings of the CIB W65 Organisation & Management of Construction,* 28 August to 3 September, Glasgow, UK, Vol. 1, pp. 572–577.

Dulaimi, M. and Langford, D. (1999) Job behaviour of construction project managers: determinants and effectiveness. *Journal of Construction Engineering and Management* ASCE, July/August, 256–264.

Dulaimi, M. and Kumaraswamy, M. (2000) Procuring for innovation: the integrating role of innovation in construction procurement. *Proceedings of the Association of Researchers in Construction Management (ARCOM),* September, Glasgow, UK, Vol. 1, pp. 303–312.

Dulaimi, M.F., Ling, F.Y.Y., and Bajracharya, A. (2002) A theoretical framework for understanding construction innovation: an organizational perspective. *Proceedings of the 10th International Symposium for CIB W55 and W65 Working Commission,* 9–13 September, Cincinnati, OH, USA, pp. 19–28.

Dulaimi, M. (2005) The challenge of customer orientation in the construction industry. *Journal of Construction Innovation,* **5**: 3–12.

Egan, J. (1998) *Rethinking Construction.* London, HMSO.

Eisenhardt, K.M. (1985) Control: organisational and economic approaches. *Management Science* **31** (2): 134–149.

Fraser, C. (2000) The influence of personal characteristics on effectiveness of construction site managers. *Construction Management and Economics* **18** (1): 29–36.

Gann, D.M. and Salter, A.J. (2000) Innovation in project-based, service-enhanced firms: the construction of complex products and systems. *Research Policy* **29**: 955–972.

Kim, W.C. and Mauborgne, R. (2005) Value innovation: a leap into the blue ocean. *Journal of Business Strategy* **26** (4): 22–28.

Kirsch, L.J. (1997) Portfolio of control modes and IS project management. *Information Systems Research* **8** (3): 215–239.

Kirsch, L.J., Sambamurthy, V., Ko, D. and Purvis, R.L. (2002) Controlling information systems development projects: the view from the client. *Management Science* **48** (4): 484–498.

Liker, J.K., Collins, P.D. and Hull, F.M. (1999) Flexibility and standardization: test of a contingency model of product design-manufacturing integration. *Journal of Product Innovation Management* **16**: 248–267.

Liu, A., and Fang, Z. (2006) A power-based leadership approach to project management. *Construction Management and Economics* **24**: 497–507.

Liu, A. and Fellows, R., and Fang, Z. (2003) The power paradigm of project leadership. *Construction Management and Economics* **21**: 819–829.

Park, M., Dulaimi, M. and Nepal, M. (2004) Dynamic modelling for construction innovation. *Journal of Management in Engineering* ASCE, **20** (4): 170–177.

Narver, J.C. and Slater, S.F. (1990) The effect of market orientation on business profitability. *Journal of Marketing* **October**: 20–35.

Ouchi, W.G. (1980) Markets, bureaucracies and clans. *Administration Science Quarterly* **25**(1): 129–141.

Slaughter, S. (1998) Models of construction innovation. *Journal of Construction Engineering and Management* **124** (3): 226–231.

Susman, G.I. and Dean, J.W. Jr. (1992) Development of a model for predicting design for manufacturability effectiveness. In: *Integrating Design and Manufacturing for Competitive Advantage* (ed G.I. Susman). New York, Oxford University Press, pp. 207–227.

Sze, E., Kumaraswamy, M., Ling, Y.Y., Dulaimi, M., Bajracharya, A. and Luk, C. (2005) Barriers to construction industry innovations—a Hong Kong perspective. *Proceedings of the Third International Conference on Innovation in Architecture, Engineering and Construction*, 15–17 June, Rotterdam, The Netherlands, Vol. 2, pp. 975–987.

15 Critical actions by clients for effective development and implementation of construction innovations

E. Sarah Slaughter and William L. Cate

15.1. Introduction

Innovation frequently occurs on construction projects, developed by all of the parties involved (Slaughter, 1993, 1998). It is the epitome of inter-organisational teams solving problems but, until recently, has not been the subject of detailed management analysis of inter-firm mechanisms for collaboration and coordination. To address this issue, recent research examined the mechanisms for inter-firm collaboration for innovation in construction, focusing explicitly on the role of the client (Cate, 2001) and on the role of the builder (Seaman, 2000). The objective of the research was to assess the applicability to construction of related research in manufacturing on inter-firm dynamics for problem solving and innovation, and to identify the major influences on the rate of successful innovation in construction projects. The research focused on specific projects and team members within those projects, and identified each innovation that was implemented on each project. The results are summarised in this chapter, focusing explicitly on 'clients driving innovation' on their construction projects.

A major challenge for clients is to establish an effective environment in which teams from multiple organisations can cooperate and coordinate their activities, particularly in response to complexity and uncertainty associated with innovation. Pinto *et al.* (1993) investigated the impacts of four factors (superordinate goals, accessibility, physical proximity and formalised rules) in achieving specific project outcomes and overall performance goals. A 'superordinate goal' transcends group boundaries through a strong and immediate commitment of the resources of more than one group to a larger purpose or objective. They found that the presence of superordinate goals increased coordination and decreased conflict among the teams, leading to improved outcomes. They also found that the remaining three factors increased the performance under certain conditions, with a somewhat weaker relationship. Chen *et al.* (1998) also found that superordinate goals can increase cooperative relationships among team members despite cultural and organisational differences.

A key factor in establishing this cooperative environment is selecting the team members and contracting for their services. Analysing the dynamics among construction teams, Nam and Tatum (1992) found that the presence of long-term relationships between the client and the team members improved cooperation, flexibility and effectiveness. In a more general study of the relationship among firms, Eisenhardt (1989) reviewed the potential relationships between the principal (i.e. client) and agents (e.g. architects, builders), and proposed that the principal can guide the performance of the agent through outcome-oriented contracts (such as fixed price bid in construction) when the principal can primarily verify the value of the outcome. However, the principal would use behaviour-based contracts (such as a cost plus fee contract in construction) when the principal can easily monitor performance to ensure it complies with the principal's objectives. Eccles (1981) found that agents would work to preserve the relationship and meet the principal's objectives when they had the prospect of future contracts. Itoh (1991) examined multi-task, multi-agent environments and found that cooperation could be increased through contract structures that provide payments contingent on other agents' performance, although Holmstrom and Milgrom (1991) emphasised that the specific tasks must be tailored to improve performance under these conditions.

Creating the appropriate team and contractual mechanisms requires a certain level of competence from the client (Estades and Ramani, 1998). The client must actively engage in the project to effectively manage the multi-organisational teams (Marshak and Radner, 1972). In addition, Bidault *et al.* (1998) found that early active involvement of all members of the multi-organisational teams can increase the generation of cost-effective ideas and mitigate risk.

The literature identified several potential mechanisms to increase the rate and effectiveness of multi-firm teams, and the research indicated that several specific mechanisms were more important relative to the role of the client in the construction setting than appeared in the manufacturing studies.

15.2. Methodology

The focus of the research summarised in this chapter was on specific innovations in particular projects, analysing the inter-firm mechanisms that influenced innovation development and successful implementation. The research was conducted at Massachusetts Institute of Technology (MIT) in the Department of Civil and Environmental Engineering, supervised by Dr Slaughter and sponsored by the Center for Innovative Product Development (a National Science Foundation Engineering Research Center).

The builder-focused analysis surveyed 29 projects constructed by 7 different construction companies (operating as either general contractor or as construction manager), and identified 50 innovations (Seaman, 2000). The projects were large, occupied facilities intended for non-industrial use, such as offices, research laboratories and educational facilities, and were located in the Northeast US to facilitate site visits and face-to-face interviews during the course of the research. The construction companies varied in terms of their geographical concentration, with two companies operating internationally, three companies operating throughout the US, and two companies operating in the New England area of the US. Primary and secondary interviews, as well as related written and graphical material, were used to compile the data on each innovation and project.

The client-focused analysis surveyed 17 projects developed by 7 different clients, and identified 67 innovations (Cate, 2001). The same type of facilities was examined (i.e. large, non-industrial, occupied buildings) and the projects were located throughout the Northern US. One client was a university, two clients were governmental institutions, one was a quasi-governmental organisation and three were large multinational corporations, including one real estate and investment firm. The same data sources and collection methods described above were employed for this study.

15.3. Results on the role of clients in innovation development and implementation

The design and construction of large-scale, complex facilities requires multiple organisations with different capabilities, responsibilities and incentives. The client must consciously and knowingly create an opportune environment to explicitly encourage the development and implementation of appropriate innovations through specific measures. Specifically, a 'smart owner' identifies the overall goals and objectives for the project, actively engages multiple organisations in an atmosphere of teamwork, establishes the relative responsibility of each team member, distributes risk and benefit through proper contract structure and delivery mechanisms and commits its own resources and organisational will to realise the objectives for the project (National Research Council, 2000).

15.3.1. Superordinate goals

The manufacturing literature on multi-team innovation strongly suggests that the presence of superordinate goals increases the incidence and rate of innovation, which is confirmed in this research. For capital facilities, examples of superordinate goals include a formal recognition of a high performance level, such as a 'green' building certification, or achievement of a strategic objective for the organisation, such as rapid facility completion to capture market share. Unlike manufacturing, however, construction offers the team members multiple avenues in which they can appropriate the benefits of the innovation(s), such as through subsequent projects with different team members, and this factor can provide unique opportunities to increase the rate of innovation in construction.

In the analysis on the role of the clients, all of the clients identified at least one and often several superordinate goals for each project, and the rate of innovation increased with the number of identified goals. While almost three quarters of the builder-analysis projects did not have an explicit overarching goal towards which all of the team members were working together, the eight projects where these builders identified one or more superordinate goals increased the rate of innovation significantly. Specifically, the average number of innovations per project was three times higher for projects with the acknowledged presence of superordinate goals than for those without the goal(s).

The disjuncture between the presence of superordinate goals acknowledged by the builder and those assumed by the client may be an indication of a lack of communication between these parties; the clients may think that the builders understand their overall goals and objectives for the project, but these goals may have been developed and expanded without the direct involvement of the builders, and may, therefore, not have been a critical influence on their priorities. In addition, almost one-third of the

Table 15.1 Reason for innovation – analysis of builders and analysis of clients

Reasons for innovation	Number of innovations (proportion)	
	Builder analysis	Client analysis
Solution	27 (54%)	42 (63%)
Opportunity	23 (46%)	25 (37%)
Total	50	67

Source: Seaman (2000) and Cate (2001).

innovations identified in the client study included management innovations, which may not have been as apparent to the builders as they were to the client.

The presence of these overarching goals and objectives for the project can change the mindset of the team members by fostering opportunities to invent or adapt novel solutions to suit the needs of the project. In the builder study, the majority of the innovations were perceived to be a solution to a specific need or challenge on the project (Table 15.1). The remaining innovations were developed to take advantage of an opportunity, such as through explicit experience with the innovation or development of particular skills and capabilities. In contrast, almost two-thirds of the innovations in the client study were seen as solutions to project needs, and only one-third were seen as an opportunity to create new benefits.

While this difference may also be influenced by the specific projects examined and the unique innovations identified in each study, it could also be due to the different ways in which the project participants can benefit from innovation. The client is primarily concerned with successfully completing the project and, therefore, is most interested in innovations that solve problems. In selected cases, the client may also be interested in testing an innovation or approach to learn whether it could solve similar problems on subsequent construction projects. In contrast, a builder is constantly assessing its competitive advantage and can actively pursue opportunities to improve its competitiveness through the successful development and implementation of innovation(s), particularly if the associated initial cost is compensated or absorbed by a current project. Recognising these incentives for innovation, it is not surprising that the builders identify a higher proportion of innovations as 'opportunistic', which could provide clients with a mechanism to encourage further innovation from builders through joint acknowledgement of potential benefits from innovation through execution on subsequent projects.

15.3.2. Team selection and organisation

Innovation, by its nature, involves some degree of risk and uncertainty. The client establishes the project context in which the project team members recognise this risk and uncertainty, and the associated costs and benefits are distributed among the parties involved. In addition, the client can select specific project team members to increase the likelihood that they will innovate. As proposed in the literature, this research found that long-term relationships among the client, designers and builders increase the likelihood of innovation. In addition, the research confirmed the initial expectation that the client

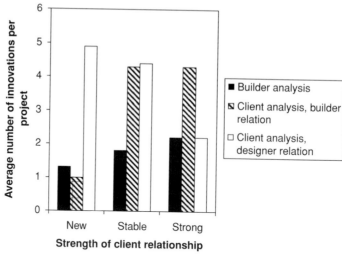

Figure 15.1   Rate of innovation by strength of relationship with client – builder and client analyses.

could rely upon outcome-based contracts (e.g. fixed price contract) in an innovative climate.

In both the builder and client studies, the average number of innovations per project increases as the strength of the relationship between the client and the builder increases (Figure 15.1). Given the uncertainty associated with innovation and the degree to which builders often assume a major portion of the cost and schedule risk associated with the execution of the project, a longer term relationship between the client and the builder can provide important insight into the manner in which each party might respond to such risk and uncertainty and enhance the trust necessary to encourage risk-taking.

However, the same pattern does not seem to apply to the designers. The client study indicates that designers are more likely to innovate in a new relationship with a client. This result is particularly surprising because the client, in most cases, can select the designers with greater latitude than it can select builders. In particular, clients can often select designers based upon their capabilities alone, while many procurement regulations require open bidding and award for the builders based on cost rather than capability assessment.

To explore this issue more deeply, the studies examined the rate of innovation relative to whether the project teams were assembled with future projects in mind. In both studies, the vast majority of project teams were not expected to work together on a subsequent project. However, the rate of innovation increased significantly for those teams that did explicitly expect to work together again (Figure 15.2). These repeat projects can significantly reduce and even eliminate the uncertainty associated with innovation for all of the team members and thereby reduce the majority of the risk, which in turn encourages innovation. In a related study, the project teams that worked together several times to implement the same innovation reported greater cost and schedule benefits with each subsequent implementation (Semlies, 1999).

Contract structure clearly defines the relative distribution of risk. As expected, the highest rate of innovation in the builder study occurred under the outcome-based

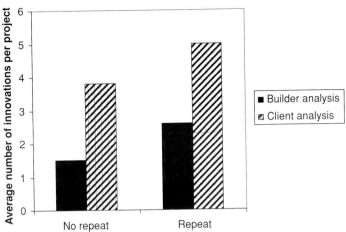

Figure 15.2   Rate of innovation by expected repeat projects – analysis of builders and analysis of clients.

contract structure (i.e. lump sum contract) for the builder, and the lowest rate of innovation occurred under the performance-based contract (i.e. fee-based compensation). Although the results for the client analysis indicate that the higher rate of innovation is associated with the guaranteed maximum price (GMP) contract type rather than the fixed price contract type, the GMP is a modified outcome-based contract that relies on verification of the quality of the outcome rather than the monitoring of performance. (However, the rate of innovation cannot be directly compared to the builder-focused analysis, since none of the client-analysis projects entailed a performance-based contract.) These results indicate, however, that projects with a clear focus on outcome with the related risks distributed to the builder increases the rate of innovation, potentially because the ultimate success of the innovation depends on the proper execution of the project, which is predominantly under the direct control of the builder.

Indeed, a direct relationship between the client and the innovating party can encourage construction innovation. In the client study, the rate of innovation was higher for projects with direct contracts between client and specialty contractor than those without such an arrangement (Figure 15.3). In a similar pattern, approximately one quarter of the projects in the builder analysis comprised direct contracts between the client and specialty designers or contractors separate from the contract with the general contractor or construction manager (Seaman, 2000). Related research in the role of designers in the generation and successful implementation of structural innovations found that the majority of the innovations were found in projects with design/build delivery mechanisms for the structural portion of the project (Semlies, 1999).

15.3.3. Client commitment

The client needs to have a reasonable degree of knowledge and competence with respect to its capital facility assets to effectively define relevant superordinate goals and to organise the design and construction team. In the client study, all of the clients had some degree of internal capability with respect to their capital facility assets, and the

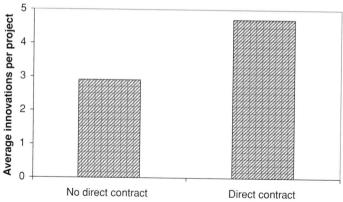

Figure 15.3   Rate of innovation by direct specialty contract – analysis of clients.

rate of innovation increased as that competence level increased (Figure 15.4). In the case of 'some' internal competence, a single programme manager was generally responsible for managing the procurement and oversight, while extensive competence involved several internal personnel, often with architecture, engineering and/or construction expertise, to manage the project.

Similarly, the commitment of the client can be measured through the extent and duration of their involvement in the project, with full commitment defined as early and uninterrupted involvement of the client throughout the entire project life cycle from planning through design and construction, and through operations and (potential) disposition of the asset. Clients with a full commitment to the project had a higher rate of innovation on their projects than clients with only partial commitment. The involvement of the client can also be measured in terms of the degree to which the client has access to the detailed records for the project, such as through audited accounts. Clients that had the right to review the accounting records for the project had a higher rate of innovation on their projects than clients who did not have such access.

Decision-making by the client, specifically with respect to the acceptability of an innovation, can be critical to ensuring that the team can sufficiently prepare, develop

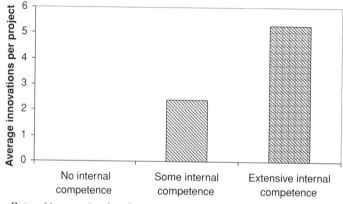

Figure 15.4   Rate of innovation by client competence – analysis of clients.

and modify its activities to accommodate the innovation. Both the builder and client studies found that the vast majority of specific innovations were identified and agreed upon in the planning process. Related research on innovations in structural systems found a similar pattern of early commitment to an innovation during the design process (Semlies, 1999).

15.4. Summary and conclusions

From the results of this research, it is clear that the client has the determining role in establishing the incidence and rate of innovation on its projects. It is the client that establishes and communicates the superordinate goals that bind the project team members together to develop and successfully implement innovative approaches. The results reveal, however, that the client needs to emphasise the nature and importance of those goals to all team members early on in the process and bridge the gap between the client, designers and builder in the recognition of those goals despite (sometimes) misaligned agendas. In addition, while the client often regards an innovation as a solution to a specific project need, builders can see opportunities for benefiting from innovation through potential application on subsequent projects. Clients can build upon these different means of appropriating the benefits of innovation to increase the builder's willingness to participate in the development and implementation of innovation.

To select and structure the team, the results indicate that the client can develop a long-term relationship with a builder to decrease the perceived risk associated with innovation and increase the rate of innovation. However, it may want to refresh its relationship with the designers to reinvigorate the development of innovative approaches. It is clear from all the studies that explicitly providing an opportunity for involvement in future projects increases the probability of innovation.

Actions such as these rest upon a strong internal competence within the client organisation, including involvement throughout the life cycle of the project and an early commitment to specific innovations despite their intrinsic risk and uncertainty. Open communication between the project team members, such as through open accounts and records, enhances the communication and trust among all parties.

As clients increase their expectations of the performance of their built facilities and face increasing cost and schedule constraints, the industry as a whole will need to develop more effective and efficient methods of delivering these assets. New pressures on long-term operating costs and environmental impacts will further drive the need for better designs and construction methods. Clients must acknowledge their pivotal responsibilities in establishing the context for their project teams to successfully develop and implement innovations to meet those needs.

References

Bidault, F., Despres, C. and Butler, C. (1998) The drivers of cooperation between buyers and suppliers for product innovation. *Research Policy* **26**: 719–732.

Cate, W.L. (2001) The client's role in inter-firm collaboration for innovation development and implementation in building construction. *MS Thesis*, Department of Civil and Environmental Engineering, Massachusetts Institute of Technology, Cambridge, MA.

Chen, C.C., Chen, X.-P. and Meindl, J.R. (1998) How can cooperation be fostered? *Academy of Management Review* **23** (2): 285–304.

Eccles, R.G. (1981). The Quasifirm in the construction industry. *Journal of Economic Behavior and Organization* **2** (4): 335–357.

Eisenhardt, K.M. (1989) Agency theory: an assessment and review. *Academy of Management Review* **14** (1): 57–74.

Estades, J. and Ramani, S.V. (1998) Technological competence and the influence of networks: a comparative analysis of new biotechnology firms in France and Britain. *Technology Analysis and Strategic Management* **10**: 483–495.

Holmstrom, B. and Milgrom, P. (1991) Multitask principal-agent analysis: incentive contracts, asset clientship, and job design. *The Journal of Law, Economics, & Organization* **7**: 24–52. [Special issue.]

Itoh, H. (1991) Incentives to help in multi-agent situations. *Econometrica* **59** (3): 611–636.

Marshak, J. and Radner, R. (1972) *Economic Theory of Teams.* New Haven, CT, Yale University Press.

Nam, C.H. and Tatum, C.B. (1992) Noncontractual methods of integration on construction projects. *Journal of Construction Engineering and Management* **118** (2): 385–398.

National Research Council. (2000) *Outsourcing Management Functions for the Acquisition of Federal Facilities.* Washington, DC, National Academy Press.

Pinto, M.B., Pinto, J.K. and Prescott, J.E. (1993) Antecedents and consequences of project team cross-functional cooperation. *Management Science* **39** (10): 1281–1295.

Seaman, R.A. (2000) Multi-organizational project teams and construction innovation: the role of the general contractor and construction manager. *MS Thesis*, Department of Civil and Environmental Engineering, Massachusetts Institute of Technology, Cambridge, MA.

Semlies, C.J. (1999) Inter-firm collaboration in the implementation of structural innovations in building construction. *MS Thesis*, Department of Civil and Environmental Engineering, Massachusetts Institute of Technology, Cambridge, MA.

Slaughter, E.S. (1993) Innovation and learning during implementation: a comparison of user and manufacturer innovations. *Research Policy* **22**: 81–95.

Slaughter, E.S. (1998) Models of construction innovation. *Journal of Construction Engineering and Management* **124** (2): 226–231.

Part 2
The innovation process

16 Overcoming resistance to innovation: the integration champion in construction

Andreas Hartmann

16.1. Introduction

In construction, it is typically the client who takes the initiative to design and construct a facility. Construction firms do not produce goods for an anonymous market, but rather provide services to specific customers whose building need triggers the construction process. Moreover, the expectations and specifications of the client strongly shape the provision of construction services and the final constructed facility (Dulaimi, 2005). As a consequence, clients also determine the possibilities for new or improved product and process solutions in construction projects. They demand particular requirements with regard to such matters as place, type, purpose and dimension of the facility or the duration of the project. It is argued that the more advanced these requirements are, the more the potential for generating new ideas increases (Gann, 2000). On the other hand, clients also directly form the way of working between project participants by, for example, applying a certain procurement strategy or choosing particular subcontractors. The more coordination, commitment and collaboration there is between project participants, so it is claimed, the greater potential there will be for implementing new ideas (Blayse and Manley, 2004; Kumaraswamy *et al.*, 2004).

Although clients driving innovation seems to be a critical antecedent for the performance of construction projects, the behaviour of construction clients towards innovation has remained vague. The research effort in this area has been low, and the existing results are inconsistent. For example, Nam and Tatum (1997) investigated ten successful innovations and found that in seven cases, a high level of client involvement in the project was critical to innovation success. Moreover, based on their findings, they identified a close relationship between the clients' technical competence and their active participation. That is, if clients are able to understand an innovative solution they are more likely to engage in and to commit themselves to the innovation. They then may take on the role of an innovation supporter in the construction project, who sponsors the new idea and protects it. The mere fact of having technical competence, however, does not necessarily mean construction clients will support the innovation. Based on three case studies on the client's role in the innovation process, Ivory (2005) concludes that clients can and do suppress innovations despite previous construction experience. In addition, he suggests that innovative construction projects led by the client are exceptions rather than the norm.

Apart from knowing that some construction clients drive, adopt or support innovation and others do not, there seems to be a considerable gap in understanding the rationale behind the attitude of construction clients towards innovation. A recent attempt to close the gap showed that several factors of the environmental, organisational and technological context account for how clients become aware of innovative solutions, how they perceive innovation attributes and how they come to an adoption decision (Hartmann *et al.*, 2006). This chapter builds on this research by arguing that successful innovation requires key individuals who can span the boundary between the client organisation and the innovative construction project and who actively promote the adoption process of the client. First, it highlights various kinds of resistance to innovation in construction and their sources. Based on that, the author then explains the role individuals may play in overcoming innovation barriers and introduces the integration champion as a key person who is able to successfully support the adoption process of construction clients. The chapter concludes with some recommendation for future research on championing in construction.

16.2. Innovation resistance

There is evidence from past research that some construction clients adopt innovative solutions whereas others do not. Moreover, the same client may implement innovations in one project but not in another. The questions remain: what accounts for the different behaviour? Why is innovation adopted in one situation whereas in another situation there is resistance to innovation?

First of all, resistance to innovative solutions is not exceptional. On the contrary, the vast majority of individuals do not seek change. According to Rogers (2003), only 2.5% of individuals in a social system have an interest in innovation for its own sake. That is not really surprising since innovations by their very nature are risky and generally complex endeavours. It is impossible to exactly define the starting problem or the final state of innovation processes as well as their social and economic consequences. Risk of failure is inherent to innovation. In addition, innovation processes do not follow fixed procedures and predetermined steps and may result in additional cost and time compared to existing practices (Hauschildt and Schewe, 2000). Given that, why should someone commit himself to an activity with an uncertain outcome and process?

16.2.1. Resistance due to risk

The perceived risks associated with innovations are major reasons for the emergence of adoption barriers. Risk avoidance, a universal phenomenon of human behaviour, may also be observed in construction. For example, the study of Ivory (2005) reveals that short-term risks such as late or over-budget projects and long-term risks such as high maintenance costs may prompt construction clients to actively slow down innovations in order to restrict possible threats to the project and the constructed facility. Here it is important to note that construction innovation acts as a multiplier of uncertainty. In construction, location and usage of a building strongly influence the design and construction processes (Reichstein *et al.*, 2005). Furthermore, construction clients must make their decision to buy on the basis of preliminary project plans

(Dulaimi, 2005). Construction projects already confront clients with uncertainties, and new constructional developments intensify these uncertainties. In particular, the fact that construction innovations can rarely be tested under a controlled environment makes it difficult to determine their long-term behaviour and their effects on other parts of the building system (Slaughter, 1998). Of course, clients can and do deal with project risks including risks associated with innovations in different ways. They can try to transfer the risks to other project participants through contractual arrangements or minimise risk through expert reports or field tests. However, in most cases, risks will remain which clients will have to accept. Moreover, even if risks are transferred, clients might be unfavourably affected due to the risk incidence. Again, why should clients engage in innovative activities? That clients are willing to bear risks at all can be attributed to the extent to which the client regards innovations to be more beneficial than standard solutions (Rogers, 2003). The greater the perceived relative advantage of an innovation is, the more likely possible risks are accepted and the innovation is adopted. However, in most cases the decision to adopt is embedded in project environments consisting of a number of stakeholders both internal and external to the client organisation who possess diverse meanings and perceptions on the risks and expected benefits of an innovation. Unless a collaborative and supportive attitude towards innovation is achieved among the different project participants, the adoption process will get stuck in heterogeneous and conflicting interests, understandings and practices.

16.2.2. Resistance due to behavioural change

Besides provoking uncertainty, innovations in most cases challenge existing technologies, products or processes. They question the predominant way of thinking and working within an organisation, and their successful implementation forces individuals to make behavioural change. As mentioned above, individuals typically tend to sustain their habits towards an existing practice and preserve the status quo rather than constantly change their behaviour. Consequently, behavioural change depicts another major source of resistance to innovation (Sheth, 1981) and whether a behavioural change can be expected depends on the degree to which an innovation is consistent with the existing values and beliefs of an individual or organisation (Rogers, 2003). Again, in construction one can also observe that habits are sustained which impede innovations and again the plurality of internal and external stakeholders may account for difficulties to attain a project environment conducive to innovation. That is, behavioural change is not a matter of a single organisation but concerns different project parties passively affected by or actively engaged in the innovation process (Slaughter, 2000). The necessity to involve, or at least to gain commitment from, several autonomous organisations with diverse knowledge domains, project interests, strategic goals and cultural orientations complicates the required behavioural change process. The behaviour of construction clients appears thus far essential as it paves the way for overcoming the resistance of other project participants. It is the client who is able to depart from conventional contractual practices and establish innovation-demanding and cooperation-encouraging procedures. However, the culture of the client organisation must allow for approaching construction projects differently and for using the potential of different procurement strategies (Hartmann, 2006). Then efforts promoting innovation can refer to existing organisational values and norms to settle the internal diversity of a client organisation.

To sum up, newness always elicits resistance due to the risks and behavioural change associated with it. In construction, resistance to innovation also occurs, and the diversity of stakeholders internal and external to project participants raises innovation barriers additionally. Thus, whether an innovation is adopted largely depends on the extent to which different internal and external project participants contribute to innovative endeavours despite their diverse perspectives and interests. Due to their prominent position in construction projects, clients may be a major source of innovation resistance as well as support. Here, the author asserts that certain key individuals have a major influence on whether a client acts as barrier or driver of innovation in a construction project. These individuals are considered essential in forming the perception within the client organisation that an innovation is beneficial and in stimulating the required behavioural change of project stakeholders to implement the innovation.

16.3. Innovation advocacy

That key individuals or champions are critical for the success of innovation processes has been emphasised by numerous researchers (Schon, 1963; Chakrabarti, 1974; Burgelman, 1983; Day, 1994; Howell and Boies, 2004). Schon (1963) formulated the credo of the research stream on championing: 'a new idea either finds a champion or dies' (p. 84).

Champions are regarded as individuals who allocate considerable effort to actively promote new ideas and to overcome organisational barriers to ensure that the ideas are implemented. They mobilise and secure resources, engage in coalition building, communicate strategic meaning around new solutions, stimulate and motivate others to support innovation (Gailbraith, 1982; Burgelman, 1983; Howell and Boies, 2004). The influence champions exert is more informal in nature than based on formal leadership or power assignment from management (Howell and Higgins, 1990). Moreover, becoming a champion seems to stem from individual motivation rather than from managerial interventions (Markham and Griffin, 1998). Champions are characterised as self-confident, persistent, energetic and risk prone (Howell and Higgins, 1990).

Since resistance to innovation may have several sources (see above), some authors argue that specific types of energy are required to support innovation processes. Different individuals may provide these types of energy and a number of innovations are not driven by just one key individual but by various champions with distinctive roles (Witte, 1977; Hauschildt and Chakrabarti, 1999). There is the power champion who, because of his hierarchical position, is able to shield innovation from opposition and to overcome unwillingness. A second key individual is the technology champion who has specific knowledge to remove barriers of ignorance. The process champion is vital for overcoming non-responsibility and indifference. He has organisational knowledge to link the people needed for the innovation (Hauschildt and Kirchmann, 2001). The fourth key person who is distinguished is the relationship champion. He or she binds people inside and especially outside the organisation (e.g. customer or supplier) to allow for inter-organisational cooperation in the innovation process (Walter, 2003). It should be noted that these roles need not necessarily be divided. They can be incorporated in one person, or individuals can change their role within and between innovation projects (Hauschildt and Schewe, 2000).

The role of champions in construction innovation has been a little-investigated research topic. Based on several case studies, Nam and Tatum (1997) claim that managers who committed themselves to new ideas were key factors for successful innovation. In addition, they suggest that these managers combined different champion roles, such as technology champion and power champion (Nam and Tatum use other terms but describe similar roles). An interesting finding is that the combination of champion roles in one person seems to be particularly beneficial for radical innovations. That contradicts other research, which emphasises the division of labour as an effect of innovation newness (Hauschildt and Chakrabarti, 1999). However, research on champions in general and in construction takes up a perspective that is mainly focused on the support of innovation within a single organisation. Little is known about supporting roles in construction projects consisting of various organisationally independent entities. It is expected that for successful innovation, individuals on the boundary between the client organisation and the innovative construction project are required who combine the characteristics of process and relationship champion. That is, people are needed who remove cooperation barriers and make an innovative idea a desired part of the project. These people can be called integration champions.

16.4. The integration champion

The assumption that integration champions will be vital for the position an organisation (e.g. the client) adopts towards innovation in a construction project stems from the innovation barriers as described earlier. Resistance to innovation is mostly caused by the perception that the benefits do not outweigh the risks of an innovation and by preference for the status quo. Construction projects may intensify resistance as they already exhibit risks and are formed by a temporary coalition of autonomous organisations. Since the client normally takes the decision to adopt an innovation and is in the position to exert influence on the behaviour of project participants, individuals spanning the boundary of client organisation and construction project are able to contribute significantly to the successful implementation of an innovative solution. If these persons actively seek to overcome the adoption barriers within the client organisation and the construction project and enthusiastically promote the innovation adoption, they will act as integration champions. More specifically, the integration champion will contribute to the innovation success by the following means.

16.4.1. Creating awareness

Typically, construction projects depend on the specific requirements of the client and the location of building. Changing demands and environmental constraints may result in constructional problems, which cannot be solved adequately with existing solutions and offer possibilities for new approaches (Nam and Tatum, 1989; Mitropoulos and Tatum, 2000; Hartmann, 2006). Integration champions raise client awareness that innovative alternatives exist which possess the potential to solve problems related to the construction project and meet client expectations more adequately than traditional solutions. They emphasise the importance construction problems have for attaining the client's project goals and show, in principle, how the innovation contributes to these goals. Creating awareness seems to be vital because construction is not the core business for a number of clients, and it is not characteristic for them to have an overview on available construction alternatives. Apart from solving problems, innovative ideas can also improve project performance by anticipating project demands and grasping technological opportunities (Nam and Tatum, 1992). The role of the integration champion here is to create a need on the part of the client and to show in principle why the innovation would make a desirable contribution to the construction project.

16.4.2. Translating innovative ideas

For the adoption of an innovation, clients must favour the new solution compared to other alternatives. They must perceive the risks as manageable and must be convinced that the advantages outweigh the disadvantages. For perception forming, it is crucial how clients interpret information about an innovation and how credible they consider the information (Rogers, 2003). Integration champions translate the information (e.g. technical specifications) into a language that is understood by the client organisation. They reduce uncertainty about the functioning and consequences of the innovative solutions by providing examples of similar solutions or the results of laboratory and field tests. Moreover, integration champions frame the innovative idea in terms of the client's strategic and cultural orientation and support a common understanding within the client's organisation about the contribution of the innovation. They shape the client's perception that the innovation is not interfering with the existing values and norms of the organisation and the prevalent way of working.

16.4.3. Encouraging communication

Innovation adoption is, first of all, a communication process through which uncertainty about a new solution is reduced and the perception of the benefits from the solution is increased. Integration champions facilitate this communication process by searching, filtering, evaluating and storing information about the client's needs and problems and about potential innovative solutions to appropriately respond to these needs and problems. They ensure that the stakeholders within the client organisation are able to understand the innovation and its contribution to the project and that the innovating firm is able to adjust a solution to the specific requirements of the client. Moreover, they transmit information on organisational, strategic and cultural issues to project participants that are relevant for developing structural and social bonds and cultivating cooperative behaviour (Walter, 1999). They identify actors that are qualified and capable to contribute to the innovation process (e.g. experts) and bring them together with other project participants. Their comprehensive network inside and outside the client organisation allows them to encourage communication by promoting contacts, coalitions and relationships (Gemünden and Walter, 1997). Building up and using such networks will help leverage innovative ideas in construction projects.

16.4.4. Coordinating activities

Besides commitment, construction innovation calls for structured and coordinated implementation. The activities to implement an innovative solution must be adjusted to other project activities without threatening the overall project goals. Particularly for the client organisation, it will be important to reduce the likelihood of any negative effects on the desired project outcomes. Here, integration champions turn an innovative idea into a plan of action (Hauschildt and Kirchmann, 2001) and coordinate the contributions of the project participants during the adoption process. They establish adequate monitoring mechanisms to detect unexpected changes and solve potential problems in time. Furthermore, they ensure that the performance of the project participants conforms to the planned innovation objectives (Walter, 1999).

16.4.5. Solving conflicts

During innovation adoption, conflicts between project participants may arise due to different views on the objectives, the available resources and the ways of implementing the innovation. Integration champions address the underlying problems of conflicts and make them visible and understandable to internal and external project participants. They identify the shared goals as well as interest differences, combine and redefine them in a way that project participants can attain their own objectives and that the overall project goals can be reached (Walter, 1999). Again the person-specific network of integration champions facilitates conflict management. Their personal and long-standing good relationships with people in the client organisation and other project participants help them to understand motivations, expectations and capabilities of the different actors involved in a construction project. Based on that, integration champions are able to establish constructive dialogues and to reduce fears of exploitation and vulnerability (Walter, 2003). Trust and commitment among project parties can evolve and complement the contractual arrangements, which are insufficient to safeguard the transaction processes in a construction project.

16.5. Conclusion and directions for future research

Innovation is often confronted with resistance. That is why innovation needs individuals who actively and enthusiastically promote new ideas and overcome barriers. This chapter argues that the existence or non-existence of such key individuals on the boundary between client organisation and construction project strongly influences the adoption behaviour of construction clients. These individuals can be called integration champions because they successfully help to embed innovation adoption by clients into the construction process. In particular, they remove cooperation barriers in construction projects and client organisations and make innovative ideas desirable for them. They are catalysers for clients driving innovation.

Although there is some evidence that integration champions exist, further research is needed to clarify their role in construction innovation and in the innovation adoption process of construction clients. Future research may focus on the question: who takes on the role of the integration champion in a construction project? Previous research showed that despite the degree of innovation, champion roles in construction merge in one person (Nam and Tatum, 1997). A single person might also act as integration champion or the role may be split between two persons; one person takes on the role of the relationship champion whereas the other person acts as the process champion. Whether the role is separated may also depend on the adoption context. In smaller organisations and straightforward projects a single person may fulfil the role. In larger organisations and complex projects a role division seems more likely. This leads to another question: on behalf of which organisation does the integration champion work? The role may evolve from the client organisation, the general contractor or architect as determined by the context. For example, the experiences and internal and external networks that professional clients can refer to will increase the likelihood that integration champions work on behalf of the client. It would also be worth studying which organisational factors are important for integration champions to arise. For instance, there is no clear evidence whether champions emerge when organisations support innovations or when they are reluctant to carry out innovations. In organisations supportive to innovations, individuals may be creative and champions may arise as a response to these favourable environments. In organisations that are more hostile to innovations, the difficulties to put an innovative idea into practice may challenge some individuals to overcome the unfavourable conditions (Markham and Griffin, 1998). However, there seems to be an association between championing and a firm's innovativeness. Within an organisation that is very committed to innovation, champions are more likely to fulfil their role. Although these organisations show well-established processes, innovation processes by their very nature are unstable and unpredictable. Champions keep the project on track by advocating the idea or finding critical resources (Markham and Griffin, 1998). If there is a considerable lack of innovation support, champions may not be able to overcome the existing barriers. As a consequence, these individuals may lose their motivation; they may reduce their level of championing or they may leave the organisation (Markham and Griffin, 1998).

Until now, championing as an aspect of construction innovation and clients driving innovation has been insufficiently researched. Expanding our knowledge of how champions emerge and promote new ideas in construction projects will be crucial for a more accelerated diffusion of construction innovation.

References

Blayse, A.M. and Manley, K. (2004) Key influences on construction innovation. *Construction Innovation* 4: 143–154.

Burgelman, R.A. (1983) A process model of internal corporate venturing in the diversified major firm., *Administrative Science Quarterly* 28: 223–244.

Chakrabarti, A.K. (1974) The role of champion in product innovation. *California Management Review* 17: 58–62.

Day, D.L. (1994) Raising radicals: different processes for championing innovative corporate ventures. *Organization Science* 5: 148–172.

Dulaimi, M.F. (2005) The challenge of customer orientation in the construction industry. *Construction Innovation* 5 (1): 3–12.

Gailbraith, J.R. (1982) Designing the innovating organization. *Organization Dynamics* 10: 5–25.

Gann, D. (2000) *Building Innovation: Complex Constructs in a Changing World*. London, Thomas Telford.

Gemünden, H.G. and Walter, A. (1997) The relationship promotor: motivator and coordinator for inter-organizational innovation cooperation. In: *Relationships and Networks in*

International Markets (eds H.G. Gemünden, T. Ritterand A. Walter). New York, Elsevier, pp. 180–197.

Hartmann, A. (2006) The context of innovation management in construction firms. *Construction Management and Economics* **24** (6): 567–578.

Hartmann, A., Dewulf, G. and Reymen, I. (2006) Understanding the innovation adoption process of construction clients. In: *Clients Driving Construction Innovation: Moving Ideas Into Practice* (eds K. Brown, K. Hampson, and P. Brandon). *2nd International Conference of the Cooperative Research Centre (CRC) for Construction Innovation*, 12–14 March, Goldcoast, Australia (on CD-ROM).

Hauschildt, J. and Chakrabarti, A. (1999) Arbeitsteilung im Innovationsmanagement. In *Promotoren—Champions Der Innovation* (eds J. Hauschildt and H.G. Gemuenden), 2nd edn., Wiesbaden, Gabler, pp. 69–87.

Hauschildt, J. and Kirchmann, E. (2001) Teamwork for innovation—the 'troika' of promoters. *R&D Management* **31** (1): 41–49.

Hauschildt, J. and Schewe, G. (2000) Gatekeeper and process promotor: key persons in agile and innovative organizations. *International Journal of Agile Management Systems* **2** (2): 96–103.

Howell, J.M. and Boies, K. (2004) Champions of technological innovation: the influence of contextual knowledge, role orientation, idea generation, and idea promotion on champion emergence. *The Leadership Quarterly* **15**: 123–143.

Howell, J.M. and Higgins, C.A. (1990) Champions of technological innovation. *Administrative Science Quarterly* **35**: 317–341.

Ivory, C. (2005) The cult of customer responsiveness: is design innovation the price of a client-focused construction industry? *Construction Management and Economics* **23**: 861–870.

Kumaraswamy, M., Love, P.E.D., Dulaimi, M. and Rahman, M. (2004) Integrating procurement and operational innovations for construction industry development. *Engineering, Construction and Architectural Management* **11** (5): 323–334.

Markham, S.K. and Griffin, A. (1998) The breakfast of champions: associations between champions and product development environments, practices and performance. *Journal of Product Innovation Management* **15**: 436–454.

Mitropoulos, P. and Tatum, C. (2000) Forces driving adoption of new information technologies. *Journal of Construction Engineering and Management* **126** (5): 340–348.

Nam, C. and Tatum, C. (1989) Toward understanding of product innovation process in construction. *Journal of Construction Engineering and Management* **115** (4): 517–534.

Nam, C. and Tatum, C. (1992) Strategies for technology push: lessons from construction innovations. *Journal of Construction Engineering and Management* **118** (3): 507–525.

Nam, C.H. and Tatum, C.B. (1997) Leaders and champions for construction innovation. *Construction Management and Economics* **15**: 259–270.

Reichstein, T., Salter, A.J. and Gann, D.M. (2005) Last among equals: a comparison of innovation in construction, service and manufacturing in the UK. *Construction Management and Economics* **23**: 631–644.

Rogers, E.M. (2003) *Diffusion of Innovation*, 5th ed. New York, Free Press.

Schon, D.A. (1963) Champions for radical new inventions. *Harvard Business Review* **41**: 77–86.

Slaughter, E.S. (1998) Models of construction innovation. *Journal of Construction Engineering and Management* **124** (3): 226–231.

Slaughter, S. (2000) Implementation of construction innovations. *Building Research & Information* **28** (1): 2–17.

Sheth, J.N. (1981) Psychology of innovation resistance: the less developed concept (ldc) in diffusion research. *Research in Marketing* **4**: 273–282.

Walter, A. (2003) Relationship-specific factors influencing supplier involvement in customer new product development. *Journal of Business Research* **56**: 721–733.

Walter, A. (1999) Relationship promoters: driving forces for successful customer relationships. *Industrial Marketing Management* **28**: 537–551.

Witte, E. (1977) Power and innovation: a two-center theory. *International Studies of Management Organization* **7**: 47–70.

17 Client-driven innovation through a requirements-oriented project process

John M. Kamara

17.1. Introduction

The importance of client involvement in the construction process has long been recognised as a key criterion for project success. This involvement includes taking an active part in defining the objectives of the project and of the organisation to deliver it, providing financing, selecting the team and exercising authority over the project organisation (e.g. Cherns and Bryant, 1984; Latham, 1994; Kometa *et al.*, 1995, etc). Clients are also now required to be more responsible in their dealings with the industry through a clients' charter, which requires participating organisations to 'commit to continually improving their performance in [four] themes of cultural change – client leadership, working in integrated teams, whole life quality and respect for people' (CCG, 2001). Support for client involvement is provided in various guidelines on how clients can best fulfil their role in the industry. These include guides for better briefing (CIB, 1997; Barrett and Stanley, 1999), setting up integrated project teams and supply chains (SFC, 2003), achieving project success (CS, 2004) and other tools, such as the design quality indicator (DQI) for defining and assessing indicators for building design quality (CIC, 2003).

The role of clients in projects, and indeed, the whole industry at large, is now seen to be more than just making sure that projects are successful. Since the realisation that practices favoured by clients (e.g. competitive lowest cost tendering) stifle innovation by encouraging an adversarial culture in the industry, it is now agreed that clients can work to promote and even drive innovation in the construction process (Latham, 1994; Egan, 1998). In addition to the demonstration projects encouraged by the Movement for Innovation (M4I), a key area that clients can make a difference is in creating enabling environments for innovation in construction projects through suitable procurement strategies. de Valence (2002) suggests that 'the move away from traditional procurement systems will have significant effects on innovation, because the traditional design-bid-build method does not allow for capture of intellectual property by construction contractors in their tenders ...' (p. 380). Such enabling environments include the management (and/or taking on of) project risks to allow a relative degree of 'freedom' by construction professionals to be innovative. An example of this is the creation and implementation of a single model environment (SME) in the Terminal 5, Heathrow Airport, UK project by BAA PLC (Lion, 2004). In this situation the client took on the risk of insuring the SME thereby making it easier for project participants to contribute to this integrated (virtual) model without fear of design liability. BAA also made a huge investment to ensure the technical feasibility of the SME (Lion, 2004).

Clients can (and to varying extents do) also contribute by encouraging the adoption of performance measurement and benchmarking (going beyond 'business-as-usual'), and the improved use of IT between the stages in project delivery and use – measures that create opportunities for innovation (Marosszeky, 1999; Manley, 2006). Performance measurement, and by implication, performance specifications, can, in particular, contribute to design creativity by removing the relative constraints that prescriptive requirements can impose on construction professionals (Gross, 1996). The National Institute of Standards and Technology (NIST) in the US, recommends that 'whenever possible, requirements should be stated in terms of performance, based upon test results for service conditions, rather than in dimensions, detailed methods or specific materials. Otherwise new materials or new assembles of common materials, which would meet construction demands satisfactorily and economically, might be restricted from use, thus obstructing progress in the industry' (qtd in Gross, 1996). Performance specifications are particularly suited to integrated procurement strategies, which are considered to provide a more enabling environment for innovation (de Valence, 2002). However, the adoption of performance specifications (and use of design-build (D&B)) depend on clarity in defining briefs and the confidence with which clients are able to monitor/ensure that performance requirements are being implemented in a project.

This chapter focuses on this aspect of client-driven innovation in construction projects that concerns the formulation, monitoring and assessment of performance-based specifications within the context of an enabling environment for innovation in projects. The assumptions on which this chapter is based are summarised as follows:

- Project innovation is enhanced by (among other strategies) an appropriate enabling environment created by the client.
- The increasing confidence with which clients can create such enabling environments partly depends on their ability to clearly articulate their requirements and the relative ease with which they can continuously monitor and assess how the process and end product of construction, satisfies their requirements.
- A way to enhance this kind of client-driven innovation is through a requirements-oriented project process (ROPP).

The chapter defines the concept of a ROPP. Strategies for the attainment and implementation of such a process are also described. These include the development of a dynamic and evolving project requirements document (PRD), and design metrics for implementing and assessing performance-based specifications in a project. The chapter concludes with a discussion on how a ROPP can enhance client-driven innovation in construction.

17.2. Requirements-oriented project process

17.2.1. Definition and rationale for a requirements-oriented project process

'A Requirements-Oriented ... [or requirements-driven] Project Process (ROPP) can be defined as a process where there is explicit ... traceability to the requirements of the client, where every action can be traced back to the original wishes of the client' (Kamara and Anumba, 2006, p. 3820). In a requirements-driven process, each activity is driven by the requirements for its execution. It is a 'customer-driven approach where recipients of information and/or resources are the "customers" of those providing that information or resource. All the parties involved in a project therefore become a network of customers and suppliers' (Kamara and Anumba, 2006, p. 3820). The ultimate goal of a ROPP is client satisfaction. The contention here is that while there are many processes that invariably lead to solutions that satisfy client requirements (and can be described as 'client-oriented'), more systematic and explicit mechanisms are required to ensure that there is continued focus on the requirements of clients. Improvements in procurement and contract strategies, the introduction of more integrated forms of procurement, improved project management practices and the special efforts that have been (and are being) made to improve the construction process over the last 10–15 years, and many other initiatives, have all contributed (and are contributing) to the delivery of better value for clients (Ndekugri and Turner, 1994; Kamara *et al.* 1996; Bennett and Jayes, 1998; PP, 2002; Perform21, 2005; etc.). However, 'because client requirements are "embedded" in the various documents and processes involved, there is a risk that client requirements become subsumed in other project requirements; exclusive focus on client requirements cannot be guaranteed' (Kamara and Anumba, 2006, p. 3822). Another concern is that contrary to the accepted wisdom that the industry has to become more client-focused (Egan, 1998), there is an emerging view that this needs to be reversed. In their book '*Why is construction so backward?*' Woudhuysen and Abley (2004) recommend that one of the ways to reverse 'backwardness' in the industry is through 'asserting the independence of architects from unbridled client power and partnership fudge [in reference to partnering]' (p. 284). Their contention is that this will give more freedom for innovation solutions to the problems of the industry. Innovated construction solutions are certainly needed, but the way to address this is not by curtailing 'client power' (as Woudhuysen and Abley, 2004 suggest), but by giving 'freedom' to designers through, for example, more performance specifications *and* also by giving clients the assurance (through objective monitoring of project processes and outcomes) that the products of industry will meet (and even exceed) their expectations. Curtailing 'client power' without these guarantees will lead to a return to the situation where designers designed for posterity without due consideration of clients' wishes. This is why an explicit ROPP is required. It is acknowledged that integrated processes such as the process protocol (PP, 2002) provide (to some extent) a framework for a requirements-oriented process. For example, one of the activity zones in the PP (Table 17.1), 'development management' is responsible for creating and maintaining business focus throughout the project. The different stages (FAVE Briefing, Mid-design, Ready for Occupation and In-use) of the DQI tool also provides a broad framework for assessing how the quality indicators defined during the FAVE (Fundamental, Added Value and Excellent) briefing stage are incorporated in the design and completed building (CIC, 2003). This chapter builds on these initiatives in developing the concept of a ROPP.

17.2.2. Objectives for a requirements-oriented project process

The concept of a ROPP is intended to make more visible the requirements of the client within a construction project. The development of a ROPP therefore starts with client

Table 17.1 Overview of the process protocol (developed from PP, 2002)

Activity zones	Process phases (0–9)									
	0	1	2	3	4	5	6	7	8	9
	Pre-project				Pre-construction			Construction		PC[a]
	Demonstrating the need	Conception of need	Outline feasibility	Substantive feasibility/outline financial authority (FA)	Outline conceptual design	Full conceptual design	Coordinated design/procurement/full FA	Production information	Construction	Operation and maintenance
Development management										
Project management										
Resource management										
Design management										
Production management										
Facilities management										
Health/safety management										
Process management										
Change management										

Source: Kamara and Anumba (2006; paper IC-595).
[a] Post-completion.

requirements – definition, capture, representation and management throughout the life cycle of a project. In this chapter, 'project process' refers to the framework for realising the vision (or requirements) of the client into a completed facility (Kamara and Anumba, 2006). A project is considered to be a unique process with start and finish dates, and consisting of a set of coordinated and controlled activities (ISO8402 in Lockyer and Gordon, 1996). A construction project is deemed to 'start' with the 'decision to build' (when it is decided by a client that a construction project is required to satisfy a particular business need), and 'end' with the completion and handover of the facility in question (i.e. the shaded area in Table 17.1 corresponding to PP phases 4–8). 'Client requirements' are defined here as the 'collective wishes, perspectives and expectations of the various components of the client body' (Kamara *et al.*, 2002, p. 3). The term 'client body' is used instead of 'client' as this includes the owner and buyer of construction services, users and other individuals or groups who influence, and/or are affected by the acquisition, use, operation and demolition of the facility (Kamara *et al.*, 2002).

The capture and representation of client requirements has been the subject of many previous research initiatives which include work by Barrett and Stanley (1999) (defining key success factors for briefing), Blyth and Worthington (2001) (various strategies for managing the brief for better design), Kamara *et al.* (1999, 2002) (on the use of structured techniques for the capture and processing of client requirements), Rezgui *et al.* (2003) (on the use of IT in managing the briefing process) and Othman *et al.* (2004) (concept of dynamic brief which links brief development to project change orders). While these initiatives have mostly focused on the initial phase of a project (although with emphasis on how briefing should integrate with design and other construction activities), it is evident that the incorporation of client requirements over the project life cycle is now being considered (e.g. DQI and work by Othman *et al.*, 2004). The main objectives of a proposed ROPP can therefore be summarised as follows:

- The capture and representation of client requirements in a 'solution-neutral' format (performance specifications).
- The progressive updating and modifying of client requirements during project implementation.
- The systematic mapping (or 'translation') of client business needs into design, construction and project management solutions throughout the project life cycle.
- The proactive management of client requirements (through traceability and compliance checking) throughout the project and facility life cycle (i.e. going beyond the end of a project into facility use and operation).

It is proposed here that the strategy to achieve a ROPP is through the development of a dynamic PRD.

17.3. Project requirements document

The idea of a PRD is similar to that for a computer-based 'Integrated Building Model', except that in the PRD, the focus is on client requirements. The PRD is conceived as a separate document with the ability to link (or integrate) it with other project documents as required. The PRD contains information about the client's wishes for the *end product* (e.g. functions and attributes about the facility) and the *process* (e.g. client's preferred level of involvement in procurement, and strategy for operation and use) to bring this about. It also contains information on all other project requirements (e.g. site and regulatory requirements) and the interrelationships between different types of requirements, with a particular emphasis on how other project requirements either enhance or hinder the achievement of client requirements. The relationship between the *process* and the *end product* should also be established (i.e. how a particular process can lead to the desired outcome). The established relationships between the different sources of information in the PRD should facilitate the planning and execution of all aspects of the project (e.g. design, project management) (Kamara and Anumba, 2006).

The development of the PRD should involve key project stakeholders and should be an evolving and dynamic document. It should also allow different end-users of this information to be able to retrieve information that is relevant to their needs, but with a link to the 'original' (or current) PRD. To facilitate this, the PRD should allow the following:

- The automatic processing, exchange (or linking) and tracking of PRD information with other project documents and/or applications (e.g. CAD systems for design).
- The continued development and modification of the PRD during the course of the project with facilities for collaborative decision-making on priorities and trade-offs within the PRD.

'Automatic processing' means that client briefing information is mapped to design metrics (defined below), and that it is possible to interrogate different sources of client/project information to aid project-related decision-making. 'Automatic exchange and tracking' means that information in the PRD is used to inform project-level actions (e.g. design, construction planning, etc.) and this information is automatically fed into key project-level processes and applications. On the other hand, project decisions/activities/actions at various stages can be checked against information in the PRD, and comparison of these with the PRD can facilitate updates or modifications in both directions (i.e. within the PRD and of project activities); in the case of project activities it can highlight non-compliance that will allow corrective action to be taken. The relationship between the PRD and project processes is illustrated in Figure 17.1. The solid lines between the PRD and project stages show the link between the PRD and each project stage (feeding into, as well as deriving from, a particular project stage). The dotted lines indicate intermediate interactions between the PRD and activities within each project phase.

17.3.1. Design metrics and the PRD

A 'metric' is a means by which something (e.g. design) is 'measured' (or assessed). It is a concept that is widely used in product design but it is being proposed here as a means to facilitate the explicit mapping of client requirements to design and construction solutions (Kamara *et al.* 1999; Kamara and Anumba, 2007). For the purpose of building, design metrics are attributes that can be used to assess whether a design/construction solution satisfies relevant client requirements. A design metric is a re-formulation of a client's 'business need' in design terms. For example, a client's need for 'adequate

External sources of information; can feed
into different versions of the PRD

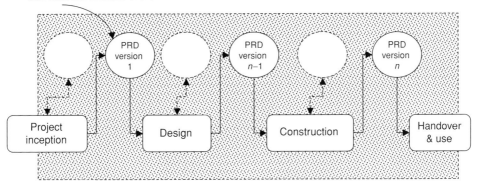

Figure 17.1   Relationship between the PRD and the project process.

space' can be expressed (in design terms) as 'gross floor area' (of say, between 400 and 500 m^2); the specific size will depend on the particular context (e.g. activities, user types, etc) of the client. 'Gross floor area' is a solution-neutral specification of the client's need for 'adequate space' and various solutions (e.g. a rectangular or octagonal room shape) can satisfy this requirement. A design metric therefore comprises of three interrelated parts:

- A statement of what is to be measured/assessed (e.g. gross floor area).
- A unit of measurement (e.g. m^2, but this need not be a quantitative measure).
- A 'target value' (e.g. a range between 400 and 500 m^2 as above). This will constitute the 'design solution space'.

The 'target values' are project-specific, but the 'statement' and 'unit of measurement' can (but not always) be generic for a variety of projects or project types.

17.3.2. Systematic mapping and traceability of requirements

The development and implementation of design metrics is central to the mapping and traceability of client requirements in the PRD. One of the key strategies proposed in the client requirements processing model (CRPM) by Kamara *et al.* (2002) is the use of the first stage of the quality function deployment (QFD) correlation matrix to map client requirements to design attributes. An extension of this over the project life cycle is illustrated in Figure 17.2. Systematic mapping is at different levels:

- from 'client's business need' to 'strategic requirements';
- 'strategic requirements' to 'design specifications' (or 'design metrics');
- 'design specifications' to 'construction specifications' (or metrics) and,
- 'construction specifications' to 'operation and use specifications'.

It should be noted that, for simplicity, only main project phases have been shown in Figure 17.2 (the 'Pre-project' stage corresponds to PP phases 0–3; 'Pre-construction

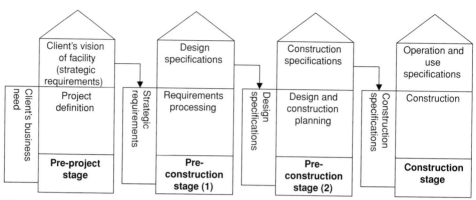

Figure 17.2 Systematic mapping of client requirements in a project. (Source: Kamara and Anumba, 2006; paper IC-595.)

stage 1' corresponds to PP phases 4 and 5; 'Pre-construction stage 2' corresponds to PP phases 6 and 7 and, 'Construction' stage corresponds to PP phase 9, in Table 17.1).

17.4. Discussion and conclusions

The central theme of this chapter is that a ROPP will contribute to enhancing client confidence in creating more enabling environments for innovation in construction projects. The 'enabling environments' considered here relate to: (a) the use of more performance-based specifications within an integrated procurement framework that will encourage more innovation in the solutions developed by construction professionals;nd (b) the support for emerging mechanisms for the live capture and reuse of project knowledge and life cycle information management. There are obviously many strategies that can motivate clients to drive through innovation in the industry as have been advocated by other researchers (e.g. Marosszeky, 1999; de Valence, 2002; Manley, 2006;. Furthermore, many procurement and contract strategies, and project systems are generally geared towards meeting the objectives of the client, and can, therefore, be described as 'client-oriented'. But since the industry has been criticised in the past for not taking client concerns seriously, making the project process more transparent and explicit through a ROPP cannot be a bad thing. In fact, it will enhance client confidence in the intentions of the industry to seek the best interests of its clients. This, it is argued, will encourage clients to be more proactive in driving innovation because the benefits from such innovation will be passed on to them (given that project processes are driven by client requirements and interests).

This chapter has also proposed that a ROPP can be achieved through the development and implementation of a dynamic PRD. It is to be noted that the PRD is very much a conceptual idea that needs further research and development. But similar ideas have already been proposed. For example, in the PP, the 'Change Management' activity (Table 17.1) is responsible for 'effectively communicating project changes to all relevant activity zones and the development and operation of a legacy archive' (PP, 2002). The legacy archive is a similar concept to the PRD. However, the difference with the PRD is that focus is on client requirements. The implementation of the PRD will also require the use of advanced information and communication technologies (ICTs) to enable the automatic processing, exchange and tracking of PRD information. This also assumes that all project information is held in electronic format in a central database, which is clearly not the case for many projects in construction. However, with the growing use of Web-hosted Project Extranets and the example of BAA in their SME for London Heathrow Airport Terminal 5 (Lion, 2004), it is feasible to assume that the PRD can be used on projects. It should be pointed out that the proposal for 'automation' in the PRD does not, and will not, exclude human involvement in the process; the intention is for it to be complimentary.

This chapter has set out proposals for enhancing client-driven innovation through a ROPP. While further research is clearly needed to realise these ideas in practice, there is every indication that their implementation will proffer benefits to all stakeholders in the construction industry.

References

Barrett, P. and Stanley, C. (1999) *Better Construction Briefing*. Oxford, UK, Blackwell Science.

Bennett, J. and Jayes, S. (1998) *The Seven Pillars of Partnering*. London, Thomas Telford.

Blyth, A. and Worthington, J. (2001) *Managing the Brief for Better Design*. London, Spon.

Cherns, A.B. and Bryant, D.T. (1984) Studying the client's role in construction management. *Construction Management and Economics* **2**: 177–184.

Construction Clients Group (CCG) (2001) *The Clients' Charter*. Available at http://www.clientsuccess.org.uk/home.html [Accessed 14th December 2006].

Construction Industry Board (CIB) (1997) *Briefing the Team*. London, Thomas Telford.

Construction Industry Council (CIC) (2003) Design Quality Indicator Online. Available at http://www.dqi.org.uk/DQI/Common/DQIOnline.pdf [Accessed December 2006].

Construction Success (CS) (2004). Achieving construction success Web site http://www.constructionsuccess.org/index.asp [Accessed December 2006].

de Valence, G. (2002) Construction innovation in globalized markets. In: *Design and Construction: Building in Value* (eds R. Best and G. de Valence). Oxford, Butterworth-Heinemann, pp. 375–386.

Egan, J. (1998) *Rethinking Construction*. London, DETR.

Gross, J.G. (1996) Developments in the application of the performance concept in building. *Proceedings of 3rd CIB-ASTM-ISO-RILEM International Symposium*, National Building Research Institute, 9–12 December, Tel-Aviv, Israel, Vol. 1, pp. 1–11. Available from: http://fire.nist.gov/bfrlpubs/ [Accessed January 2007].

Kamara, J.M., Anumba, C.J. and Evbuomwan, N.F.O. (1996) *Integration of Design and Construction: A Review of Existing Approaches*. Technical Report No. 96/2, School of Science and Technology, University of Teesside, UK.

Kamara J.M., Anumba C.J. and Evbuomwan N.F.O. (1999) Client requirements processing in construction—a new approach using QFD. *Journal of Architectural Engineering* **5** (1): 8–15.

Kamara, J.M., Anumba, C.J. and Evbuomwan, N.F.O. (2002) *Capturing Client Requirements in Construction Projects*. London, Thomas Telford.

Kamara, J.M. and Anumba, C.J. (2006) Specifications for a requirements-centred project process. *Proceedings of the Joint International Conference on Computing and Decision Making in Civil and Building Engineering*, 14–16 June, Montreal, Canada (CD-ROM), pp. 3817–3828.

Kamara, J.M. and Anumba, C.J. (2007) The 'voice of the client' within a concurrent engineering design context. In: *Concurrent Engineering in Construction Projects* (eds C.J. Anumba, J.M. Kamara and A-F. Cutting-Decelle). Abingdon, Taylor and Francis, pp. 57–79.

Kometa, S.T., Olomolaiye, P.O. and Harris, F.C. (1995) An evaluation of clients' needs and responsibilities in the construction process. *Engineering, Construction and Architectural Management* **2** (1): 57–76.

Latham, M. (1994) *Constructing the Team: Final Report on Joint Review of Procurement and Contractual Arrangements in the UK Construction Industry.* July, HMSO, London.

Lion, R. (2004) Terminal 5 single model environment—vision, reality and results. *Design Productivity Journal* **3** (3): 53–58.

Lockyer, K. and Gordon, J. (1996) *Project Management and Project Network Techniques*, 6th edn. London, Pitman.

Manley, K. (2006) Identifying the determinants of construction innovation. In *Proceedings of the Joint International Conference on Construction Culture, Innovation and Management (CCIM)* (ed. M. Dulaimi), 26–29 November, Dubai, UAE. [Paper 160 in CD-ROM Proceedings.]

Marosszeky, M. (1999) Technology and innovation. In *Building in Value: Pre-design Issues* (eds R. Best and G. de Valence). London, Arnold, pp. 342–355.

Ndekugri, I. and Turner, A. (1994) Building procurement by design and build approach. *Journal of Construction Engineering and Management* **120** (2): 243—256.

Othman, A.A.E., Hassan, T.M. and Pasquire, C.L. (2004) Drivers for dynamic brief development in construction. *Engineering Construction and Architectural Management* **11** (4): 248–258.

Perform21 (2005) *Partnering Contracts.* Available from Perform21 Web site at: http://www.perform21.com/contracts/ [Accessed February 2006].

Process Protocol (PP) (2002) *Process Protocol.* Available at: http://www.processprotocol.com/ [Accessed February 2006].

Rezgui, Y., Bouchlaghem, N.M., Austin, S.A. and Barrett, P. (2003) 'An IT-based approach to managing the construction brief'. *Information Technology in Architecture, Engineering and Construction* **1** (1): 25–38.

Strategic Forum for Construction (SFC) (2003) Strategic Forum Integration Toolkit. Available at: http://www.strategicforum.org.uk [Accessed December 2006].

Woudhuysen, J. and Abley, I. (2004) *Why is Construction so Backward?* Chichester, UK, Wiley-Academy.

18 Knowledge management supports clients driving innovation: two case studies

Marja Naaranoja, Päivi Haapalainen and Heikki Lonka

18.1. Introduction

The construction sector is traditionally seen as fragmented, which in itself can be a real barrier to innovation. In construction there are a few large companies that have resources for Research & Development (R&D) functions. Small and medium-sized enterprises (SMEs) in construction tend to specialise, competing with each other in their chosen field, usually on the basis of price, due to most clients' emphasis on the lowest initial costs. Innovation is therefore a low priority, representing a gamble to SMEs who tend to offer what they know they can deliver in a competitive, local market place.

However, innovation is required as a means to develop better quality buildings or other products that will fulfil the needs of future users. Further management innovations mark a departure from traditional construction management principles, processes and practices or a departure from customary organisational forms that significantly alters the way the work of the management is performed. According to Hamel (2006), the key challenge seems to be generating truly unique ideas. He proposes that four components can help: a significant problem that demands fresh thinking, creative principles or paradigms that can reveal new approaches; an evaluation of the conventions that constrain novel thinking and, examples and analogies that help redefine what can be done. This chapter studies knowledge management (KM) in two construction projects and how KM supports management innovation or innovation in general. A well-planned KM system itself can be seen as an innovation in the construction industry, where typically knowledge is managed in the same way as it has been done in the past. Unfortunately, this way is seldom innovative.

This chapter focuses on KM and KM-related innovations by exploring offered, needed and wanted knowledge in construction projects. It explores what kinds of issues are critical for clients to succeed in driving innovation. However, it does not discuss the criteria for success. Instead it tries to find out how knowledge resources are critical to innovation and why they are accessed or not accessed by clients in general.

18.2. Knowledge management

The term 'knowledge management' refers to the creation, codification and dissemination of information in relation to a wide range of knowledge-intensive tasks (Harris *et al.*, 1998). KM should begin with recognition of the kinds of knowledge that require management; in addition, it requires recognition that any new knowledge that is needed must be created and used by the right people.

Much of the information that is available to an organisation has an independent existence that can be shared in the form of news, entertainment and education (Lillrank, 2002). Any data that have some value for someone in some context can be processed into information. This 'information' becomes a component of 'knowledge' when it is critically analysed such that its underlying structure is understood.

According to Argote and Ingram (2000), knowledge is embedded in the three basic elements of organisations – members, tools and tasks – and the various sub-networks formed by combining or crossing the basic elements. Quinn *et al.* (1996) divided the knowledge of an organisation into four levels:

(1) know what: the cognitive knowledge, know what knowledge is basic knowledge that individuals can acquire through extensive training;
(2) know how: the ability to translate bookish (know what) knowledge into real world results, know how is capabilities and skills that are informal and hard to describe precisely;
(3) know why: the ability to take the know-how to unknown interactions, it is deep knowledge of cause-and-effect relationships and,
(4) care why: self motivated creativity, this level of knowledge exists in a organisation's culture, care why knowledge enables organisations to simultaneously thrive in the face of today's rapid changes and renew their cognitive knowledge, advanced skills and systems understanding in order to compete in the next wave of advances.

18.3. Method

The research is qualitative in nature aiming at illustrating the challenges of KM in construction projects. This chapter is based on the literature review and the evidence arising from two case studies in Finnish towns. The case studies are: (a) a vision building process in a school building project and (b) the role of user representative in another school building project. In both cases the old buildings were renovated and some new facilities were built. The first case, vision-building process, was action research. The aim was to improve the communication within the project and vision building was introduced as an innovative way of improving the knowledge sharing in the project. The second project was to observe and interview one of the teachers (end user representative) who took an important role in the construction process. She actually became an expert in construction renovation and her role was acknowledged in the project team.

Theme interviews were used as a means to collect information but artefacts (e.g. project drawings, memos) were also collected from the two cases studies; and, observations were also made of meetings held in two of the projects. The themes used in the interviews were: the phases of the project; participation methods in these phases; the main information brokers in each phase; experienced successes and challenges and, organisation of the information and knowledge exchange methods. In addition, the interviewees were encouraged to raise any issues that they thought would be important for additional understanding. Case materials were scanned in order to find out what kind of knowledge was used and what the informants said about the knowledge types by using the Table 18.1 as a framework.

18.4. Case study 1 – vision building

The vision-building process was created in cooperation with the project organisation of a school project. The vision building had three objectives: to motivate stakeholders to take part in the design process (especially school personnel), to ensure that the whole team (architect, other designers, school representatives, political decision makers and authorities of the different departments of the town, project manager and facility department) shared the same view of the project and its objectives and to use the vision-building process as a tool for prioritising the different needs and wishes of the stakeholders. The vision building involved over 70 stakeholders, without a well-defined practice many voices would not have been heard.

18.4.1. Vision building in practice

A half-day session was organised by the researchers and city personnel to build the shared vision for the renovation project. The parties invited to the session included all the staff from the school (including kitchen and cleaning staff), some parents of the students, representatives from the school office and city planning office, building and maintenance offices from the city, museum office and all the designers involved in the project that had been already chosen. Altogether 69 people participated in the vision-building process. The largest group of the participants represented the people working at the school. The building department of the city was represented by several people, including the project manager, the architect and supervisors.

Before the session some of the participants had been asked to write a short introduction about their own viewpoint and goals for the project. After the deputy mayor of the city had opened the session these introductions were presented and 12 of the following viewpoints were given. The representatives of school office and town planning office talked about future student numbers in different areas of the city and about developing the teaching. The head master of the school presented some of the problems at the school and the wishes of the teachers for the renovation. The project manager introduced some of the limitations for the project that budget and timetable would encounter. The representatives of the maintenance department and other service groups (e.g. cleaning staff) of the city introduced the needs of maintenance. The architect spoke about the kind of information the designers needed for the design process.

Altogether 12 different viewpoints were presented. It was clear that these introductions helped the different parties to understand that their own needs and wishes were not the only ones and that it was unlikely to be able to make all the different wishes

Table 18.1 Different types of knowledge (and resources required) in project-based industry (Järvinen, 1999)

Types of resources	Types of knowledge				
	Conceptualised knowledge	**Operationalised knowledge**	**Cultural knowledge**	**Grounded knowledge**	**Coded knowledge**
Long-lasting physical resources				Products, prototypes, process technology	
Human resources (individual)	Facts, concepts, principles etc.	Know-how tacit knowledge	Behavioural models, values		
Human resources (community)	Collective beliefs	Cooperation and communication	Values, goals, ideologies, etc.	Roles, routines, rituals	
Knowledge and information base resources					Web pages, databases, manuals etc.

come true. The introductions also showed that, in addition to the small core team of the project, there were several different parties involved in the project.

Participants were divided randomly into small groups consisting of seven to eight people to discuss the kind of school that they wanted with a facilitator group made up of the representatives of the different parties. Three different scenarios were given to the groups: a small village school, scenario of poorness (sticking to a strict budget) and a specialised school. The aim of this was to ensure that the groups considered different possible futures and to keep the vision at an abstract level. In the small village school scenario the idea was that the school was the heart of the area and that lots of other activities other than teaching happened at the school. The scenario of poorness brought consideration of the fact that the economical situation was getting tighter and that education for a large group of students must be provided with only a little money. In the specialised school scenario consideration was given to the fact that competition between schools could increase in the future and schools would have to specialise in order to attract students and to be able to survive.

Participants in each group considered what the most important things about their future school were. Each person in the group had to come up with his/her own ideas and wrote these down on paper. Using the ideas generated, the group voted for the three most important factors.

Ideas developed in the groups were then presented to other groups by writing them on paper and hanging the papers on the wall. Words like 'safety', 'cosines', 'practical' and 'good acoustics' came up. After the presentations all participants cast their votes for the three most important factors. The eight factors that got the most of the votes were as follows: practical (21 votes), safety (19 votes), facilities that are versatile and can be changed according to needs and thus support specialisation (13 votes), a construction process is safe and of good quality (12 votes), versatility (9 votes), a school that emphasised culture and presentation (8 votes), economy in the whole life cycle (7 votes) and 'student must come first' (7 votes).

Next a common vision was created by discussion, based on the voting results. Discussion about the actual meaning of the different words emerged and some changes were made so that the vision would better serve its future purposes. The final vision was as follows:

> A practical and safe school that supports specialisation by multipurpose rooms and will be economical during its whole life cycle as well as providing good quality.

Many participants believed that there was a need for this kind of session. For example, all the teachers of the school participated in the session in their own time. People were also quite eager to prepare and present the introductions when they were asked to do so. The feedback from the session was very positive; people felt that they had had a chance to influence and also to learn some information. It was even said that vision building should be done in every construction project.

The objectives of the vision building were all met. All participants committed to making the vision come true and the project manager took the major responsibility for this. The feeling of having a chance to influence particularly motivated the teachers to participate in the project. Later during the project the vision was used as a guideline of what the goal of the project was when deciding which of the end user wishes should be realised and which were not so important.

Christenson and Walker (2004, p. 43) claim that merely preparing the vision is not enough. There also has to be a communication strategy that helps to share the vision in the organisation. The process of togetherness created the vision. It definitely acted as a process of communicating the vision. However, the negative side of the process was the costs. Bringing together a large group of people for half a day to undertake vision building cost a lot. The vision building process also facilitated learning in the project. The presentations, particularly, gave people a chance to learn other parties' viewpoints.

The vision building process can be seen as a management innovation since it is used now in the new projects and the end users claimed that it really helped them.

18.5. Case study 2: the role of user representative

Case study 2 was a very difficult school renovation project. During the design phase the budget had to be doubled, because the condition of the structures was not as good as had been expected. More surprises were revealed during the first phase of construction. The outside wall was wet; there was no insulation in the foundations and the concrete floor was completely wet from underneath. Even the interior walls were damaged by moisture from cleaning water. It was soon found out that the whole structure should have been studied more carefully during the programming phase. The case study illustrates the role of the art teacher in the problem solving and how she participated in the process.

The architect and the art teacher had already taken part in the actual design and construction of the school about 30 years ago. It was natural, therefore, that the art teacher continued to be the school's representative in the design team. The school took the renovation project very seriously and invested extra resources in it by giving the art teacher resources. The art teacher was nominated as a part-time team member and in addition the rest of the staff actively participated. This was not common in city school projects. Normally it was the principal who would attend the design and construction site meetings.

At the beginning the school had already realised that it had to take the project seriously. All the important plan solutions and designs were made together at the school. The architect and the art teacher were the core team members during the programming and throughout the project. The art teacher was also seen as a mediator of information:

> It is important that there is one person who possesses all the knowledge. The art teacher has mediated the design team's ideas to the teachers and vice versa.
>
> – the principal

During the design stage there was confusion about who was leading the project; as nobody but the art teacher took the leading role. Most people thought that the city office architect should have been the project manager, because he was the chairperson of the design meetings. He, however, saw chairing as his only role in the project.

During the construction stage the construction project manager took the leading role. During this stage the relationship between the school and the city architect's office got extremely difficult when mould was detected and analysed in a laboratory. The city architect's office refused to give the results to the school. The art teacher had to contact the laboratory to get the information. It was the art teacher who started to read about the mould problems. It seemed nobody had anything against this. 'The art teacher has carried the most painful burden' was the principal's interpretation. 'It is easier to do things when somebody else takes responsibility. The junior and the senior construction manager willingly listened to the art teacher's expertise and questioned her, a teacher who knows about bacteria. This saves their honour'.

Here we see transactive memory in action (Moreland *et al.*, 1996; Moreland, 1999;. A new question arose and the team had to decide who would take care of it and acquire more knowledge. The art teacher became an expert in these matters, because she was the most suitable person and the others had their hands full of everyday tasks. The institutional roles had no place here: it was a matter of who was the fittest for the task at hand.

This case study shows that in some situations the role of a customer or user representative can change to the role of a construction renovation expert. However, one can claim that this would not have happened if the project had been a normal project and if the user representative had not had thorough knowledge of that building and construction project in that town. However, the construction team was clever enough to accept the unusual involvement of the art teacher in the project throughout the project. The art teacher had 'care why' knowledge and she had also started to learn things consciously to also get to learn 'what knowledge'. She also disseminated 'what knowledge' in the team and 'know how knowledge' was created together in the construction team.

The described role of the art teacher was not planned and it is difficult to propose that a user representative will take on this type of role in future projects and thus the role of the user representative is not an innovation in itself. However, this case study shows that user representatives may carry an important role in a construction project if he/she is given autonomy to do that and the construction professionals let it happen.

18.6. Knowledge creation and learning

Knowledge is one of the most important assets in innovations. The big question is, as stated by Davenport and Prusak (1998), how do organisations manage what they know? KM should naturally begin with realising what kind of knowledge is there to be managed. In this regard, Järvinen (1999) proposed that five different types of knowledge play a part: (i) conceptualised knowledge, (ii) operationalised knowledge, (iii) cultural knowledge, (iv) grounded knowledge and (v) coded knowledge (see Table 18.1). The same author also defined the kinds of resources required for these different types of knowledge in an organisation (see Table 18.1).

The first category of knowledge, *conceptualised knowledge*, consists of the facts, concepts and principles involved with a given project (Järvinen, 1999). When this conceptualised knowledge (including collective and individual beliefs about how a project should be managed) is put into practice it becomes the second form of knowledge, *operationalised knowledge*; such operationalised knowledge determines behaviour – that is, it determines how people actually operate. This includes details on how client needs are found. The third type of knowledge, *cultural knowledge*, consists of the values and habits of stakeholders (Järvinen, 1999). This plays an important role in knowledge creation in a project context; every project manager needs to learn about the culture of his or her clients and stakeholders. The fourth category of knowledge, *grounded knowledge*, is embedded in the real physical objects involved in the project, and is also found in the roles, routines and rituals of the organisation. The fifth type of knowledge, *coded knowledge*, is found in documents, agreements, Web systems and other databases. Clients need to make decisions typically based on coded knowledge – the drawings and plans.

Järvinen (1999) further argued that the types of knowledge resources that are required in an organisation are: (i) long-lasting physical resources, (ii) human resources (individual and communal) and (iii) a knowledge and information base. The role of resources becomes increasingly important as the KM of a project is enhanced. These types of knowledge are studied in detail later in the case study discussion.

One of the challenges presented by KM is getting people to share the various categories of knowledge they possess and the resources they control. In some organisations, the sharing of knowledge is natural and easy; however, in others, especially in projects, the sharing of knowledge can be more difficult. This is often due to the major characteristics of construction business: uniqueness (among client representatives there are novice participants), complexity (the client needs several experts to solve the problems) and discontinuity (the construction professional can seldom work with the same client).

Explicit knowledge is formal and systematic. It can be expressed in words and numbers and can be easily communicated and shared. Tacit knowledge is subjective and intuitive and thus difficult to process or to be transmitted (Barney, 2002). The client may have difficulties in delivering the knowledge on what are the most important factors in the building. The process of turning either tacit or explicit knowledge into tacit knowledge means learning. It is important to acknowledge that participants of the project team are different (they have different backgrounds, educations and experiences); and, possess different kinds of knowledge (this knowledge is often such that the participants themselves do not even know that they have it, it is so called 'tacit knowledge').

Transferring tacit knowledge between people is not easy and even when the tacit knowledge is turned into explicit knowledge (the client representative – the art teacher in case study 2 – was able to tell that something was wrong in the existing building but not able express what it was), it may not be understood and thus cannot be used by other participants. Combining explicit knowledge with another piece of explicit knowledge may create new knowledge under favourable circumstances. And then the circle begins again: learning, sharing, creating new knowledge. And all these processes are needed in organisations especially when innovations are developed.

Miller (2003, p. 15) argues that 'there appears to be no one formal or structured process that an organisation can adopt to lead it along the pathway to successful workplace learning and to become a learning organisation'. In practice, several different methods are applied from very strict formal ways, like educational courses to informal, almost 'invisible' ways, like communities of practice (e.g. Brown and Duguid, 1991; Lave and Wenger, 1991). However, Garvin (1994) has introduced three stages that must be gone through if learning is to be useful. The first stage is cognitive, in which new ideas are welcomed as well as new ways of thinking. The second stage is behavioural change that means that the new ideas of the first stage are learnt and used in practice, in this case, the use of vision building. The third stage is the effect of the two earlier stages into the organisation – the new rule on how visions are generated in future in the organisation. This kind of a new rule was set in collaboration in the case of town organisation.

18.7. Available and used knowledge resources

The knowledge resources in the cases are studied in the following order: (1) long lasting physical, (2) knowledge and information based, (3) individual and (4) human community resources.

18.7.1. Long-lasting physical resources

The grounded knowledge of construction projects can be found, for example, in products. The window, for example, is not designed each time from first principles; they can be selected from widely available lists. In the renovated buildings there is considerable knowledge already grounded in them. The construction team needs to learn how it is built, maybe sometimes also why it exists as it currently does. Innovations might take place if the need is so unique that the existing products do not fulfil the requirement. In our two case studies there were no such unique needs.

18.7.2. Knowledge and information-based resources

Knowledge and information-based resources are the type of knowledge that is possible to be delivered in written format. It is often called coded knowledge. In practice it may be manuals, drawings, text books, papers, Web pages and other computer systems. The aim of the vision building was to produce a new and better guideline to find the shared goal of the project. In the beginning there was no guideline to create a vision for the project. There was not a description in the case projects of what was the role of the user representatives in the projects; or a clear decision-making system. In the case study 2 that enabled the user representative to take a leading role in the project, this was not usual in normal projects.

18.7.3. Human resources (individual)

Each party played his/her individual role in the project based on his/her knowledge on cooperation and on the construction process. For example, in the second case study, the experienced art teacher took an expert role in the project and she even partly led the project. However, the end users seldom have any previous knowledge on construction projects and, therefore, they have much to learn. The first case, vision building, turned out to be a powerful learning tool for the users of the multiple perspectives of the project, even the construction professionals were able to learn in that process. The professionals particularly created some kind of pre-understanding from their past experience concerning how things must happen in the project.

A knowledge structure is a mental template that individuals impose on an information environment to give it form and meaning. There are basically two fundamentally different ways of structuring knowledge: theory driven or data driven. For example, a construction specialist might think that in a certain region all the buildings need piling. Data-driven thinking requires data on the site before such a piling decision can be made. Theory-driven thinking helps individuals to understand, but it also uses stereotypic thinking controlled information processing; filling data gaps with typical but perhaps inaccurate information. This approach prompts one to ignore discrepant and possibly important information of the existing knowledge structure, and inhibits creative problem solving (Walsh, 1995). In the vision-building process the needs of different stakeholders were forced to be heard. The construction professionals had doubts about the value of the process. At the end of the project some of them still had doubts and they tried to prove that the process was not necessary and that it was even harmful by causing extra costs for the project.

An individual often works in a routine way and by this he/she avoids evaluating the meaning of every fact. We do not even use our existing knowledge in an optimal way. The user representative in case study 2 had many ideas of how things should be. She needed to find external experts that supported her opinions. She even got information about the mould problems etc. In a hasty situation this might not happen and the internal expert might either change his/her opinion without negotiating or just leave it as it is, knowing that the solution will not last.

18.7.4. Human resources (community)

Collective beliefs rule our behaviour. One example of such behaviour is the belief that an architect is not able to think economically. As a matter of fact, we assume knowledge of the values, goals and ideologies of our partners even though every party is individual. However, it is important that cultural knowledge is understood.

Organisational knowledge structures rely on consensus and agreement (Weick, 1979). This can be seen as shared frames of reference, recall of past events, the creation of events, the creation of stories and myths, vicarious learning, unlearning and memories (Lyles and Shwenk, 1992). When the vision-building process was formulated the four researchers and three practitioners exchanged past experiences and so were able to decide what kind of process was best.

Socio-political themes, such as credibility and power, influence the acceptance of the view (Mintzberg, 1983). It is understandable that an experienced expert can select rather freely his perception to the usage of new knowledge like computer systems. People allow them to select their opinions freely. Inexperienced ones have to show first how much they understand before they are heard. This kind of thinking – who knows what – is accepted and the people who have either formal or informal power can easily reject new ways of working or new knowledge – if they stated that this is not reliable the rest of the construction team believes it. The key decision maker in the use of vision-building process was the deputy mayor of the town. One can even claim that the vision building would not have been realised without his power. However, just after this project the end users of another facility arranged another vision-building meeting in order to specify the shared goal for that building project. That decision was made without the power of the deputy mayor.

Initially the users of the building communicated their knowledge to the professionals. Although users of the building were involved in the discussion, it was the norm that their views were often ignored and the manager took complete control of defining their needs.

At the beginning of the design process people who will utilise the building communicate their knowledge to the professionals. How much the professionals are able to learn and understand is questionable. Sometimes it might also happen that the power filter effect is on who is asked – this means that we only ask the people who have power but not the opinion of shop level workers. In case study 2, the user representative did not have any power filters and she asked every person who used the building about their needs. It is difficult to study all the ideas coming from the workforce in a workplace. In the vision-building process almost every stakeholder had the possibility of telling their opinions. The question of future needs was given in the first case study at the beginning of the process but afterwards questions about future needs were not regularly asked since the professionals were afraid of design changes. The following three examples of the KM processes give an idea as to what kinds of possibilities are there:

- *Mock up room.* Different groups of the end users and clients are able to study a mock up room and test how it functions. The comments can be either gathered or the whole group can join in testing the use of the space. These kinds of rooms were found to be most helpful.
- *Activity cards.* These describe in written format the functions of each space, connection between the spaces and other information, for example, how many people would use the space normally and the maximum it would hold. The cards help the client and users to describe what kind of qualities required by the space are most important to them. The architect could then read the cards and discuss the ideas raised with the user groups later.
- *Information and communications technology (ICT) tools.* For example, virtual models of the building in order to make the design solutions visible.

When the designers study the building with the users the cost expert should also be involved. Since new ideas often raise costs before further developing the ideas there should be some kind of critical checkpoint with the whole project group in order to avoid last minute quick decisions.

18.8. Conclusions

It is clear that innovations are needed in the construction industry. This chapter shows that KM innovations, new ways of doing things (vision building, increased customer autonomy), can lead to satisfied customers. The study revealed many sources of knowledge and that it is impossible to utilise all the available resources. The study also showed that the knowledge resources are difficult to utilise since we as human beings are not able to change our perceptions easily but also that the human communities might prevent the fast changes – like the professionals who did not believe that the vision building gave any real input for the process though the client and the end users were really happy that their voice was heard and that they were able to learn in the vision-building process.

Construction projects incorporate many people and at best cooperation can inspire better solutions but at worst we can hurt the feelings of other people and that might affect their motivation. Individuals can learn to think in two ways, data driven and theory driven. The data-driven approach helps us to see things in new ways and listen to other parties – especially the professionals who should learn to listen with open minds to the opinions of the client and the end users of the building. The theory driven-approach understands the phenomenon via theories and makes it possible to understand quickly, without all the facts available but misunderstanding is, of course, also possible.

By sharing knowledge with the parties it is possible to improve the KM and gain new innovative solutions to the old problems. Pre-understanding of how things must happen can be shared among the teams. Working alone would not provide us the chance to develop new knowledge. The vision building demonstrated that knowledge sharing was crucial to develop better ways of doing things – to enable the client and end users to understand the different viewpoints of the parties involved.

No formal process is available that ensures the needed utilisation of the knowledge resources. However, by giving the client and his customers, end user representatives autonomy it is possible to provide new information for the construction team. If you work alone your mental models are not tested as often – the art teacher had the role of regularly testing the solutions of the construction team and it is believed that her role was important in that case study. If the question of future needs is not examined the planned building may be fit only for the current situation despite understanding at the beginning of the project that the long-term needs are also important.

References

Argote, L. and Ingram, P. (2000) Knowledge transfer: a basis for competitive advantage in firms. *Organizational Behaviour and Human Decision Processes* **May 82** (1): 150–169. Available at http://www.idealibrary.com

Barney, J.B. (2002) *Gaining and Sustaining Competitive Advantage*, 2nd edn. Englewood Cliffs, NJ, Prentice-Hall.

Brown, J.S. and Duguid, P. (1991) Organizational learning and communities of practice: toward a unified view of working, learning, and innovation. *Organizational Science* **2** (1): 40–57.

Christenson, D. and Walker, D.H.T. (2004) Understanding the role of vision in project success. *Project Management Journal* **35** (3): 39–52.

Davenport, T.H. and Prusak, L. (1998) *Working Knowledge: How Organizations Manage What They Know*. Boston, MA, Harvard Business School Press.

Garvin, D. (1994) Building a learning organization. *Harvard Business Review* **71**: 78–91.

Hamel, G. (2006) The why, what and how of management innovation. *Harvard Business Review*, **84** (2): 72–84.

Harris, K., Fleming, M., Hunter, R., Rosser, B. and Cushman, A. (1998) *The Knowledge Management Scenario: Trends and Directions for 1998–2003*. Technical Report, Gartner Group.

Järvinen, A. (1999) Facilitating knowledge processing in a workplace setting. In: *Proceedings of Researching Work and Learning Conference* (eds K. Forrester, et al.), University of Leeds, 10–12 Sept, 677–682.

Lave, J. and Wenger, E. (1991) *Situated Learning: Legitimate Peripheral Participation*. Cambridge, Cambridge University Press.

Lillrank, P. (2002) The quality of information. *International Journal of Quality & Reliability Management* **20** (6): 691–703.

Lyles, M.A. and Shwenk, C.R. (1992) Top management, strategy and organisational knowledge structures. *Journal of management studies* **22** (2): 155–171.

Miller, P. (2003) Workplace learning by action learning: a practical example. *The Journal of Workplace Learning* **15** (1): 14–23.

Mintzberg, H. (1983) *Structure in Fives: Designing Effective Organizations.* New Jersey, Prentice-Hall.

Moreland, R.L., Argote, L. and Krishnan, R. (1996) Socially shared cognition at work. In: *What is Social About Socially Shared Cognition?* (eds J.L. Nye and A.M. Brower). Thousand Oaks, Sage.

Moreland, R.L. (1999) Transactive memory: learning who knows what in organizations. In: *Shared Cognition in Organizations: The Management of Knowledge* (eds L.L. Thompson, J.M. Levine and D.M. Messick).. Mahwah, NJ, Erlbaum, pp. 3–31.

Quinn, J., Anderson, P. and Finkelstein, S. (1996) Managing professional intellect: making the most of the best. *Harvard Business Review* **March–April**: 71–80.

Walsh, J.P. (1995) Managerial and organizational cognition: notes from a trip to memory. *Organization Science* **6** (3): 280–321

Weick, K.E. (1979) *The Social Psychology of Organizing.* Reading, MA, Addison Wesley.

19 Implementing innovations in infrastructures for the built environment: the role of project developers, customers and users

Marcela Miozzo and Nuno Gil

19.1. Introduction

This chapter explores the factors underscoring the attractiveness for, and resistance of, project developers, customers and users to the introduction of new technologies in developing new infrastructures for the built environment. We draw on a multiple case study research of the introduction of six technologies across four project settings in a large engineering programme to add a new terminal campus to one of the world's busiest international airports. Of particular interest to our research is how the plurality of stakeholders, especially the project developers, customers and end-users, affects the process of adoption of new technologies.

We argue that two major factors influence how developers, customers and users resist or embrace new technologies in infrastructures for the built environment: complexity of the selection environment and technological complexity. To develop this argument, this chapter is organised as follows. Firstly, we review the literature on innovation, especially that on the role of users in innovation and modularisation of product design. We then discuss our research method. The following section draws on concepts from the innovation literature to examine the empirical evidence of stakeholders' stance in relation to the new technologies. Finally, we discuss implications for theory and practice.

19.2. Background

The point of departure for this research is the work of Nelson and Winter (1977), which defines the 'selection environment' as the set of conditions favouring the development and adoption of an innovation. It is the environment into which the innovation is launched – which is sometimes the market – that determines whether or not the inno-vation will be a success (ibid.). The main elements of the selection environment include, first, the value of an innovation, or, stated differently, the analysis of the expected ben-efits and costs; second, the investment and imitation processes, that is the expansion path of the innovating firms and the diffusion process to imitators; and third, the way in which consumer preferences and government legislation and regulations influence what is profitable. In non-market selection environments, such as innovations in the school system, the separation of interests between firms and customers is not as sharp as in pure market environments (ibid.). These concepts are useful for analysing the determinants of innovation in the built environment as they stress both the importance of meeting consumer needs for the success of an innovation as well as the role of the different actors in shaping the development and use of a technology.

Our research also draws on the work on production and innovation of complex products and systems (CoPS) (Hobday, 1998; Miller *et al.*, 1995; special issue of *Research Policy* in 2000), which has mainly focused on the suitable organisational forms for producing CoPS. This literature explores the project and other management capabilities necessary for the organisations engaged in projects (Davies and Hobday, 2005; Gann and Salter, 2000; Brusoni *et al.*, 2001). Our research setting, an airport, is a particular case of Hobday's (1998) broader category of CoPS. In common with other CoPS, airports are not mass-produced for final customers but involve high-value capital goods produced in a one-off project manner for intermediate customers. As with other CoPS, intermediate customers are involved in innovation. Airports are physically very large and expensive developments, involving major capital investments. They are an important hub of economic activity, which can assume a significant role in the local and national economy. The recent privatisation of major airports in Europe also manifests an important socio-economic evolution from a time when physical infrastructures were owned by public authorities to the present days when infrastructures are becoming regulated businesses.

Another set of literature that informs our research is the work on the role of users in innovation (von Hippel, 1976, 1988), which explores the role of these stakeholders in technology adoption. The work by von Hippel and colleagues found that users play a dominant role in innovation in several sectors, including pultrusion process, scientific instruments and semiconductor and electronic subassembly manufacturing equipment (e.g. von Hippel, 1976, 1977, 1988). This research shows that much of the learning that can benefit a manufacturer's product development efforts may lie in the consumer environment. It also shows that it is very difficult to produce accurate judgements regarding user needs for new and 'high tech' products. Innovations triggered by users and manufacturers tend to differ, with users tending to develop 'functionally novel' innovations rather than quality improvements (Riggs and von Hippel, 1994). Von Hippel (1986) recommends manufacturers to engage in marketing research that focuses on 'lead users' of a product or process. 'Lead users' present strong needs, which may become widespread in the near future; can be a source of important benefits to many other users, and can provide valuable design data to manufacturers (Urban and von Hippel, 1988). Von Hippel and colleagues also advocate the development of 'user toolkits for innovation' (Thomke and von Hippel, 2002; von Hippel and Katz, 2002). These toolkits enable consumers to take part in need-related product development, allowing them to design and re-design the products or processes as they go through a learning process.

Finally, another stream of research that informs our work is on the nature of product design architecture. Modular product designs exhibit a one-to-one mapping from functional elements to physical subsystems and components, as well as decoupled interfaces between the elements (Ulrich, 1995). Modular designs by definition exhibit

Table 19.1   Summary of the cases

Project setting	Case	Technology adopted
Baggage handling system	IT technology for baggage reconciliation	Yes
	RFID technology for baggage identification	No
Inter-terminal train Airfield	Concrete technology for airfield pavements	Yes
	CCTV technology for vehicle occupancy Security	No
Terminal buildings	Wireless technology for way finding	No
	IRIS technology for immigration control	Yes

built-in options: the standard interfaces allow the modules to evolve parametrically as long as changes conform to the design rules and integration protocols are agreed upfront (Baldwin and Clark, 2000, p. 223). Option-holders can exercise the options by substituting one module with a superior module, and by adding new modules. The cost of modularising an integral design, however, can occasionally be unaffordable when developers need to search in a complex solution space for ways to decouple the functional elements and to integrate them economically later (Baldwin and Clark, 2000). In these circumstances, an overly high investment on modularisation may involve costly cycling behaviour that may not pay off through corresponding gains in performance improvement (Ethiraj and Levinthal, 2004). Modular design architectures encourage, however, the formation of a community of user designers.

19.3. Methodology

We base our analysis on an exploratory multiple case study research (Eisenhardt, 1989; Yin, 1994). We examine processes to develop and implement six new technologies across four infrastructure projects forming part of a large capital programme to expand an airport system. We studied the process of adoption of the following six new technologies in an airport: IT technology for reconciliating baggage, radio frequency identification (RFID) technology for baggage tracking, new concrete mix technology for airfield pavements, closed circuit television (CCTV) technology for vehicle occupancy security, wireless technology for displaying way – finding information and iris-recognition technology (IRIS) for automating immigration control. Some of these technologies were adopted by the development teams, while others were rejected upfront or abandoned mid-course (see Table 19.1). The complexity of the selection environment varies from low to high. Some technologies interact in a modular fashion with other IT and building subsystems, whereas other technologies exhibit integral or hybrid interactions. We developed our analysis by iteratively playing empirical data against concepts in the innovation literature. We resorted to tabular cross-comparisons to test the plausibility of the concepts (Miles and Huberman, 1994).

The field study took place from May 2004 to January 2007. When we set off to collect data, the project designs were substantially developed, and the physical execution stages were progressing concurrently with the design detailing stage. Data collection involved: (i) 39 face-to-face interviews, lasting from one to two-and-a-half hours with representatives of the project developer, customers and suppliers; and (ii) analysis of over one hundred archival documents, including project drawings and specifications, design standards, project briefs, and clips from the trade and business press. All interviews were tape recorded, transcribed, and organised into a digital database.

Our research setting shows the difficulties in defining the 'client' in infrastructures for the built environment. The airport expansion programme included various projects such as new terminal buildings, aircraft stands, and an inter-terminal train system. The airport was owned and operated by a subsidiary business (Airport Ltd) of a publicly listed company (Airport PLC). Airport PLC set up an independent business unit that acted as the overarching developer of the expansion programme and managed the whole programme. This unit, in turn, was formed by more than ten distinct project development units (project developers). Some of the project customers were business units of Airport PLC (e.g. retail division) and of Airport Ltd (e.g. building operations, security). Other project customers were the airlines (who would move to the new terminal campus in the future) or the immigration authority (who would operate the immigration hall). Some of the end-users were staff employed by the project customers, while others were the passengers and the retail tenants.

19.4. Implementing innovations in the built environment: empirical evidence

19.4.1. Complexity of the selection environment

We start the empirical analysis by investigating the complexity of the selection environment for adopting and implementing the six technologies. We characterise the selection environment along two dimensions: the plurality of institutional stakeholders involved in the adoption of a new technology, and the need to reconcile stakeholders' assessments of the perceived costs, benefits and operational risks in adopting and implementing a new technology.

19.4.2. Plurality of institutional stakeholders

Our fieldwork systematically uncovered a number of institutional stakeholders involved in the process of agreeing and authorising the adoption of a new technology. These invariably included one or more development teams working for the airport operator (project developers), the airline and/or other customers of the projects, statutory authorities and other third-party stakeholders (see Table 19.2).

In the case of the baggage handling system, for example, the respective development team was funding most of the capital investment. Some parts of the project scope, however, tapped into the airline's budget as in the case of baggage tracking and

Table 19.2 Developers, customers, end-users and regulatory framework of the technology

Case	Developers and customers	End-users	Regulatory framework
IT technology for baggage reconciliation	*Many* Developers: baggage project development team, airline Customers: airline, Airport Ltd (baggage operations unit)	*Many* Baggage handlers (unionised); other staff of the airline and airport	*Few* Mostly an internal business operation
RFID technology for baggage identification	*Many* Developers: baggage project development team, airline Customers: airline, Airport Ltd (baggage operations unit)	*Many* Check-in staff (airline); security staff (airport)	*Some* National standards available, international ones to be developed as technology is taken up
Concrete technology for airfield pavements	*Many* Developers: airfield project development team Customers: airline, air traffic controller, Airport Ltd (airfield operations unit)	*Many* Pilot crews (airline); ground staff (airport)	*Heavily regulated* Aviation regulator sets national safety standards, plus international standards
CCTV technology for vehicle occupancy Security	*Many* Developers: train and IT systems development teams Customers: Airport Ltd (various units such as security, building operations, IT systems)	*Many* Security staff (airport), IT systems staff (airport)	*Some* Security standards set by Department for Transport
Wireless technology for way finding	*Many* Developers: development teams (building, way-finding, IT) Customers: Airport Ltd (various business units, such as building operations, retail, security)	*Many* Passengers, airport staff	*Hardly any* Mostly an internal business operation
IRIS technology for immigration control	*Many* Developers: immigration authority, development teams (building, IT), police, intelligence agencies Customers: border control team, Airport Ltd, airline	*Many* airport operations division, passengers, immigration authority staff	*Heavily regulated* Part of large-scale border modernisation programme

reconciliation. Hence, the airport operator and airline assumed joint responsibility for the development of the baggage handling system. The airline was responsible upstream for developing the baggage check-in (involving all weighing and identification of baggage) and downstream for baggage reconciliation. The airport operator was responsible for other parts of the project scope, including baggage screening, transportation, sorting and unloading.

Likewise, the iris-recognition technology involved various stakeholders. The design of the 5 month trial in 2002, for example, was run in conjunction with two airlines, the airport, a technology supplier and the immigration authority. The airlines helped to identify approximately 200 passengers who travelled frequently; these passengers had to be approved by the border control authority. The Airport Ltd provided special areas in an existing building to install the automated barriers. Passengers had to agree to look into a video camera, which took a close-up image of the iris, and let the authorities hold the data in their computerised system. These efforts were orchestrated with the help of International Air Transport Association's Simplifying Passenger Travel Interest Group. We discuss next how the plurality of stakeholders leads to heterogeneous interests and the need to orchestrate a system to reconcile these.

19.4.3. Reconciliation of assessments of costs and benefits

The cases showed that for a technology to succeed, each stakeholder must perceive individually that its assumed benefits (which include potential savings and improvements in functional value) outweigh, first, the cost to progress with the adoption of the new technology; and second, the operational risks stemming from implementing a new technology. The cases in which new technologies were successfully adopted repeatedly suggest situations where stakeholders managed to orchestrate a system that enabled them to agree collectively that each one would invest some time, effort and capital to roll out a new technology. In contrast, the cases in which new technologies were ruled out or abandoned mid-course show stakeholders failed to reconcile the heterogeneities between their interests and assessments.

In one group of cases, as for example with IRIS, the orchestration of interests across stakeholders materialised from the outset. The airline wanted to be renowned for taking a lead in technological advances. Airport Ltd was interested in looking at new and innovative ways of improving passenger journeys through the airport. The immigration authority, in turn, was interested in working in partnership with other stakeholders to make the most of science and improve passenger clearance at immigration control. Admittedly, the costs of developing and implementing the technology drew primarily from the immigration authority's budget. Still, the expected benefits to the other stakeholders were necessary and sufficient to offset the expenditures associated with their own contributions in kind and perceived risks.

In a second group of cases, however, the orchestration of interests across stakeholders did not materialise. While a number of developers and/or customers could be very keen to adopt the new technology, others reasoned that the assumed benefits did not outweigh the costs and risks. The airport expansion business unit, for example, let the airline lead the development process of the baggage handling system for the first 2 years, but then demanded back its control as it got increasingly concerned with volatility in the airline requirements and risk of cost escalation. The airline was interested in specifying a sophisticated baggage handling system that could automatically deliver baggage from check-in point to the jetty to meet operational targets in aircraft turn-around. Such a system ought to effectively communicate with the airline systems of flight dispatch and display, baggage information displays and security systems. Stated differently, the airline was interested in being a 'lead user' (von Hippel, 1986) in baggage handling. In contrast, the airport's baggage handling development team was interested in specifying a less high-tech, more affordable and reliable system, involving both manual and automated operations. Unlike the airline, the airport was not interested in being a 'lead user' in this domain.

19.4.4. Technological complexity

We characterise the technological complexity along two dimensions. First, we examine the familiarity of stakeholders with the technologies, and the extent to which they were competent to specify their needs. Second, we examine the design architecture of the new technologies, particularly the modularity of the new technology (see Table 19.3).

19.4.5. 'Novelty' of the technology

Some technologies considered by stakeholders during project delivery were new to them in the sense that they manifested state-of-the-art applications of novel (IRIS) or more mature (RFID, CCTV) scientific knowledge in an airport environment. In these cases, the new technologies had not yet been fully developed and proven, and technology adoption involved developing new software coding and undertaking extensive trials. This novelty, compounded with lack of regulation, generated a generalised perception of 'complexity'. Specifically, novelty led to perceptions of: (i) high risks that the technologies would fail to operate reliably; (ii) high costs to develop and implement the technology; (iii) high costs to maintain the technology in the future; and (iv) difficulties to procure technology competitively. Admittedly, stakeholders found it hard to quantify objectively these costs and benefits because of their limited know-how. This (lack of) understanding increased their reluctance to support a new technology, especially if they needed to spend money from their own budget.

The 'novelty' of the technologies could also make it difficult for development teams, and stakeholders in general, to specify the technology requirements because they lacked the capabilities and know-how to do so. The baggage handling development team, for example, explicitly stated that they would not adopt RFID tagging until they could learn from other airports how to specify the requirements accurately.

Likewise, the inter-terminal development team acknowledged that until they consulted with the Department of Transport in 2004, they were unable to appreciate the requirement involved in threat *assessment*. All they had in the brief was that they should maintain segregation between departing and arriving passengers and objects (for the purpose of threat *detection*). This team subsequently abandoned the CCTV-based solution after it realised that it could not afford to pay for the development costs necessary to develop a threat assessment system.

Table 19.3 Novelty and design architecture of the technology

Case	Novelty of the technology	Design architecture of the new technology		
		Modularity	Interfaces	Functional mapping
IT technology for baggage reconciliation	*Low* Application of mature programming know-how	Yes	Standard software interfaces, plus modular components (handhelds, PCs, workstations)	One-to-one: reconciliation technology needed to identify bags going into the containers
RFID technology for baggage identification	*High* No proven application in airport systems	In part (hybrid)	International standard IT interfaces were not yet agreed; modular hardware (tags, readers, workstations)	One-to-one: possible, but there was ongoing R&D work to agree other functions for the RFID tags
Concrete technology for airfield pavements	*Moderate* High-performing mixes available elsewhere (e.g., bridges)	Yes	The scientific know-how of concrete mixes exists beyond the pavements where it gets applied	One-to-one: concrete mixes are needed to build airfield pavements
CCTV technology for vehicle occupancy Security	*Moderate* New application of mature CCTV technology; new algorithms needed	Yes	Modular software and video cameras, but cables integral to train cars	One-to-one: technology needed to detect and assess threats.
Wireless technology for way finding	*Low* Trivial application of mature technology	Yes	Modular software and components	One-to-one: wireless signs convey way-finding information to passengers
IRIS technology for immigration control	*High* New application of fresh science (algorithms)	In part (hybrid)	Modular software, but physically coupled automated barriers	One-to-one: possible, but there was ongoing R&D to find range of functions for the new technology

19.4.6. Design architecture of the new technology

Our findings suggest that a modular interaction does not suffice per se to persuade one or more stakeholders to invest in a new technology. Nevertheless, design modularity may moderate the resistance of stakeholders to new technologies in the sense that modular designs make it possible to keep a number of development options open at limited cost. Both the development team and the airline, for example, were interested in adopting a new baggage reconciliation system. The decision to migrate over 400 baggage handlers to a new system, however, was a difficult one for the airline. It involved consulting and negotiating with the union the training requirements to counteract industrial relations issues later on. The modular interaction between the two systems made it possible for the development team to wait over 3 years until the airline finally committed to fund the new reconciliation system, at low risk of delaying the overall project.

Likewise, the most expensive components, the CCTV-based solution for the vehicle occupancy security system (software, CCTV cameras), interacted in a modular fashion with other IT and physical subsystems (metal skin of the train cars). Developers managed to postpone efficiently the procurement of these elements, albeit agreeing to install the wiring looms when the fabrication of the cars started in 2003. The cables were integral to the train cars because they run between the inner and outer skins. As a result, developers incurred limited sunk costs after they decided to abandon the CCTV-based solution in 2005, while leaving the option open for the future. As put by the design manager (2005): 'Perhaps CCTV technology can evolve in the way baggage-screening machines have and then it will allow detecting threat and making threat assessment. If that happens, the cable looms are already there'.

Modular systems can also be attractive because they offer flexibility to accommodate evolution in design requirements economically. The developers of the reconciliation system, for example, structured a database with the flexibility to support a number of assumed job roles in the baggage handling teams. They also developed a tool sitting on top of the database to give these end-users flexibility to extract tailored reports from the system. This 'toolkit' (von Hippel and Katz, 2002; von Hippel and von Krogh, 2003) saved them the cost of developing reports when end-users had still limited information about their needs. Less spending upfront logically increased the attractiveness of the technology.

19.5. Implications

Our research combines concepts from different strands of the innovation literature to investigate key factors affecting the adoption of new technologies by developers, customers and users in the development of infrastructures for the built environment. We suggest that two major factors play a crucial role in this process: complexity of the selection environment and design complexity of the technologies.

The selection environment for the adoption of technologies in infrastructures for the built environment is very complex (Nelson and Winter, 1977). There is a multiplicity of institutional stakeholders involved in agreeing and authorising the adoption of new technologies. This includes developers, customers and end-users of the technology, all of which have a stake in the introduction of the new technologies. Also, adoption of most of the technologies taps into the budgets of more than one stakeholder. The new technologies in many cases are also subjected to a host of standards and regulations from national and international bodies, and even police and intelligence agencies. This heterogeneity of stakeholders' interests means that stakeholders need to reconcile their interests and their different assessment of costs (of development and implementation), risks and expected benefits (savings and functionality improvements). The groups of stakeholders that manage to orchestrate timely a system that enable them to agree on individual investment of time, effort and capital tend to succeed in the adoption of new technologies.

Second, the design of the technologies themselves is complex, in many cases involving technologies that are not fully developed or proven. This is not unusual for CoPS (Hobday, 1998), but in a context of complex regulatory framework, including security standards and regulations, stakeholders may adopt a more prudent stance. Design complexity makes it difficult to both quantify objectively the costs and benefits of a new technology and specify the technology requirements. Our results also show that the design architecture of the new technologies is important. Design modularity may moderate the resistance of stakeholders to new technologies. It is long known that modularity helps developers operationalise some strategic option-like thinking, thereby building flexibility to accommodate economically changes in design requirements (Baldwin and Clark, 2000; Smit and Trigeorgis, 2004). On one hand, it enables developers to stagger the capital costs sunk in a development effort, keeping open the option to invest more if uncertainties resolve favourable later on (Smit and Trigeorgis, 2004). On the other hand, modularity helps developers keep open the option to abandon with limited sunk costs if uncertainties resolve unfavourably (ibid.).

Our research findings have implications for practitioners and policymakers involved in the development of physical infrastructures, such as airports. First, our research highlights the complexities of the selection environment in the development of infrastructure, especially the plurality of stakeholders and heterogeneity in expectations of costs, risks and benefits associated with a new technology. Unlike the final users of commercial goods, CoPS stakeholders tend to be prudent in the use of their budgets and risk averse in terms of performance reliability. Accordingly, promoters of the new technologies should develop innovation toolkits specifically aimed at helping stakeholders understand better the new technology and reconcile conflicting interests. Second, design modularity does not automatically guarantee adoption of new technologies in complex environments, but may make it easier to orchestrate an agreement among stakeholders. Practitioners and policymakers may want to consider supporting efforts to develop new technologies with modular architectures. The modularisation of the interfaces between a new technology and other subsystems may be costly, but it may be a way to incorporate the needs of the stakeholders that otherwise would be resistant to new technologies.

References

Baldwin, C.Y. and Clark, K.B. (2000) *Design Rules: The Power of Modularity*, Vol. 1. Cambridge, MA, MIT Press.

Brusoni, S., Prencipe, A. and Pavitt, K. (2001) Knowledge specialization, organizational coupling, and the boundaries of the firm: why do firms know more than they make?. *Administrative Science Quarterly* **26** (4): 597–621.

Davies, A. and Hobday, M. (2005) *The Business of Projects: Managing Innovation in Complex Products and Systems.* Cambridge, Cambridge University Press.

Eisenhardt, K.M. (1989) Building theories from case study research. *Academy of Management Review* **14**: 532–550.

Ethiraj, S.K. and Levinthal, D. (2004) Modularity and innovation in complex systems. *Management Science* **50** (2): 159–173.

Gann, D. and Salter, A. (2000) Innovation in project-based, service-enhanced firms: the construction of complex products and systems. *Research Policy* **29** (7–8): 955–972.

Hobday, M. (1998) Product complexity, innovation and industrial organization. *Research Policy* **26**: 689–710.

Miles, M. and Huberman, M. (1994) *Qualitative Data Analysis: An Expanded Sourcebook.* Thousand Oaks, CA, Sage.

Miller, R., Hobday, M., Leroux-Demers, T. and Olleros, X. (1995) Innovation in complex systems industries: the case of flight simulation. *Industrial and Corporate Change* **4**: 363–400.

Nelson, R.R. and Winter, S.G. (1977) In search of useful theory of innovation. *Research Policy* **6**: 36–76.

Riggs, W. and von Hippel, E. (1994) The impact of scientific and commercial values on the sources of scientific instrument innovation. *Research Policy* **23**: 459–469.

Thomke, S. and von Hippel, E. (2002) Customers as innovators: a new way to create value. *Harvard Business Review* **80** (4): 5–11.

Smit, H.T. and Trigeorgis, L. (2004) *Strategic Investment: Real Options and Games.* Princeton, NJ, Princeton University Press.

Ulrich, K. (1995) The role of product architecture in the manufacturing firm. *Research Policy* **24**: 419–440.

Urban, G. and von Hippel, E. (1988) Lead user analyses for the development of new industrial products. *Management Science* **34** (5): 569–582.

Von Hippel, E. (1976) The dominant role of users in the scientific instrument innovation process. *Research Policy* **5** (3): 212–239.

Von Hippel, E. (1977) The dominant role of user in semiconductor and electronic subassembly process innovation. *IEE Transaction on Engineering Management* **2**: 60–71.

Von Hippel, E. (1986) Lead users: a source of novel product concepts. *Management Science* **32** (7): 791–805.

Von Hippel, E. (1988) *The Sources of Innovation.* New York, Oxford University Press.

Von Hippel and Katz, R. (2002) Sifting innovation to users via toolkits. *Management Science* **48**: 821–833

Von Hippel, E. and von Krogh, G. (2003) Open source software and the 'private- collective' innovation model: issues for organizing science. *Organization Science* **14** (2): 208–223.

Yin, R.K. (1994) *Case Study Research: Design and Methods*, 2nd edn. Newbury Park, CA, Sage.

Part 3
Moving ideas into practice

20 Client driven performance improvement strategies for the construction industry: development and implementation challenges

Aminah Robinson Fayek, Jeff H. Rankin and Ernie Tromposch

20.1. Introduction

Innovation in the construction industry takes various forms, including supporting performance improvement strategies and developing and implementing best practices. This chapter describes the work of two organisations in Canada that are helping to deliver innovation to the construction industry, the Canadian Construction Innovation Council (CCIC) and the Construction Owners Association of Alberta (COAA). The chapter begins by setting the context of the topic of innovation within the construction industry, with emphasis placed on the role of clients in the process. A discussion of the activities of the CCIC to support the measurement of performance in the industry, and of the COAA to develop industry best practices follows. This chapter also illustrates how industry expertise from both of these organisations is being combined with university-based research to help deliver innovation to the Canadian construction industry.

20.2. Innovation in the construction industry and the client's role

Many practitioners and researchers alike agree that the architectural engineering and construction industry can improve its overall performance (measured in terms of cost, time, safety, quality, sustainability, etc.) by creating a better business environment that encourages innovation (Manseau and Seaden, 2001). Innovation is defined in this context as the 'application of technology that is new to an organisation and that significantly improves the design and construction of a living space by decreasing installed cost, increasing installed performance and/or improving the business process (e.g. reduces lead time or increases flexibility)' (Tooles, 1998, p. 323). The innovation process can be viewed as the transition from something that is new to something that has become standard practice (Tornatzky and Fleischer, 1990), as shown in Figure 20.1. This progression involves a number of steps that can be analysed from a macro level by defining and measuring industry level metrics (e.g. percentage of firms adopting a specific innovation), or at a micro level by examining specific innovation and user level metrics

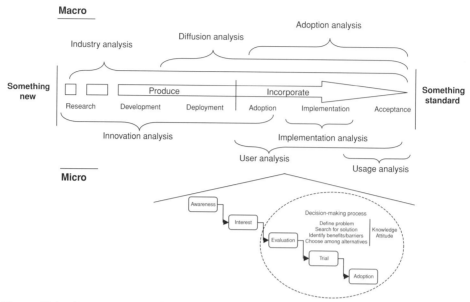

Figure 20.1 A process view of innovation.

(e.g. the complexity of an innovation or the business objectives of a firm) (Rankin and Luther, 2006). At the micro level, a number of steps are involved in adopting an innovation, as depicted in the decision-making sub-process shown in Figure 20.1 (Rogers, 1995). Essential to this discussion is that the processes are highly dependent on the effective management of knowledge.

A system of innovation contains a key catalytic role identified as the innovation broker. This role can be a focal point for a variety of parties in the construction industry (e.g. researchers, solution developers, industry associations, contractors, designers and construction owners), and facilitates the management of knowledge within an industry through its cyclical brokering activities. These types of activities have evolved to include multiple stakeholders (e.g. contractors, designers and owners), because the innovations of today and tomorrow require participation of multiple parties within the supply chain, extending across traditional organisational boundaries.

The construction client (typically the owner) is essential in many ways to a successful process of innovation. Their role includes the ability to reduce risks associated with trying something new (e.g. alternative contracting strategies), and to establish objectives that focus on long-term gains (performance) versus short-term gains (operational). Furthermore, the client has the ability to provide access to information for measurement and assessment, and to mandate innovation and change within the industry. To ultimately effect a positive change in the performance of the industry by enhancing innovation, the concept of continuous improvement is relevant. A basic maturity model progresses through the following levels: (1) standardisation, (2) ability to measure, (3) ability to control, and finally (4) ability to support improvement. Although typically applied to individual organisations, this model is equally applicable to an industry, as are the basic techniques of measurement, assessment and knowledge management. The remainder of this chapter discusses two such initiatives undertaken in Canada with the key support of construction clients and aimed at facilitating innovation to the construction industry.

20.3. Canadian construction innovation council pilot project on benchmarking

The CCIC is a not-for-profit national organisation dedicated to improving the construction industry through the increased adoption of innovative processes and technologies. The CCIC's mission is to identify needs and opportunities for industry related research and innovation, develop national innovation priorities, set targets for assessing industry performance, promote technology transfer and champion innovation issues affecting the institutional, commercial and industrial sectors. The CCIC has recognised a lack of suitable performance measurement in the Canadian construction industry and has established an Industry Performance Benchmarking Committee to: (1) measure the performance of the Canadian construction industry against a variety of parameters that will give a more complete picture of its performance; (2) provide a process that can be repeated to give an indication of change in the performance of the industry from year to year; and (3) provide data that can be used to compare the Canadian construction industry to that of other countries.

The CCIC has initiated a pilot study to test the selected performance metrics in the Canadian construction industry, and to learn what refinements are required prior to launching a full performance benchmarking study across Canada. The unique aspects of this pilot project in relation to the theme of 'clients driving innovation' include the primary partners, and the efforts in extending the definition of performance. The primary partners for this project are public owners, including two large municipalities and a national organisation responsible for the capital infrastructure of the national armed forces. The performance metrics include not only quantitative measures such as cost, time, scope, safety and quality, but also more qualitative measures such as innovation and sustainability. Although project level metrics are being gathered, it will be necessary to aggregate them to provide performance indications at the organisation and industry level.

For the pilot study, 37 public sector projects (where private sector contractors are often involved) have been studied across Canada targeting two main sectors: buildings and water/waste water pipes. The Hole School of Construction Engineering at the University of Alberta (Edmonton, Canada) was involved in the data collection and analysis process for the pilot study. As anticipated, some of the challenges experienced in the pilot study stem from differences in terminology underlying the performance metrics and the organisational differences from project to project, which impact the format in which information is captured (and hence retrieved). The results of the pilot study have been used to assess the validity of the metrics chosen, the appropriateness of the data collection methodology, and potential methods of data analysis and presentation to industry stakeholders. Based on the results, the metrics used and the data collection process were successful. The owners found that the metrics were useful, easily understood and provided valuable information. While the cost, time, scope and safety information is readily available, as expected, the information for quality, innovation and sustainability is not or requires more detailed in-person interviews of project participants. The data analysis and presentation conducted to date indicates that a combination of radar charts and box plots works best as a means to communicate results. A full-scale benchmarking program for the Canadian construction industry is currently under development, with the goal of ongoing collection of data and reporting of results.

20.4. Construction Owners Association of Alberta best practices

A central mechanism to enhance the opportunity for innovation in the construction industry is the development and implementation of best practices related to project planning, management and execution. A best practice can be defined as 'a technique or methodology that, through experience and research, has reliably led to a desired or optimum result' (*Webster's New Millennium*™ *Dictionary of English*, 2006). Best practices in construction can be implemented by many different parties involved in the execution of a project, although they are often most effective on a widespread basis if they are client mandated and driven.

A good example of client driven best practices are those developed by the COAA, a leading group of owners representing many sectors of the Alberta industrial construction community. The COAA, founded in 1973, is an Alberta-based organisation whose mission is 'to provide leadership to enable their owner members to be successful in their drive for safe, effective and productive project execution by

- providing a forum for owner members to collaborate;
- creating and promoting best practices;
- serving as the owners' voice to audiences that can make a difference;
- providing a forum for dialogue and debate among owners, EPC (engineering, procurement, construction) firms, labour providers and governments; and,
- bringing new ideas to industry and government leaders.' (www.coaa.ab.ca)

The vision of the COAA is to address the challenges facing the Alberta construction industry and to promote industrial construction excellence in Alberta. Their membership consists of 25 Principal Members, who are primarily construction owners and users of construction services, and 70 Associate Members, who are primarily engineering consultants and contractors involved in the execution of construction projects. The COAA also has affiliations with various construction industry associations, such as labour and safety associations. The COAA's governance comes from a board of directors, consisting of 12 members, each of which must be from a principal member.

Having identified a need to address significant issues facing the construction industry, the COAA has established a number of best practices committees, whose goal is to 'develop, document and facilitate the implementation of construction best practices that improve the construction industry's project performance in safety, quality, cost and schedule' (www.coaa.ab.ca). Their efforts at delivering best practices are structured around five primary areas: safety, workforce development, contracts, rework and construction industry performance, each of which is represented by a best practices committee and sub-committees. Committee representation consists of a mix of owners, contractors, engineering consultants, labour organisation representatives, government representatives and academics. Committee work is supported by over 200 representatives, providing in excess of 7500 person hours of effort per year. A significant number of

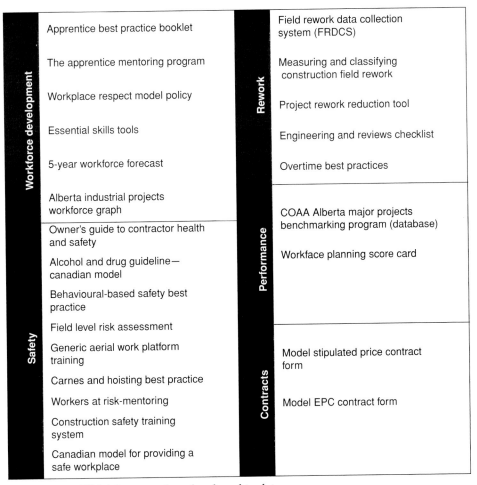

Figure 20.2   COAA best practices developed to date.

best practices have been developed to date by the COAA, as shown in Figure 20.2, and grouped according to each of the five best practices committees. Further information on each of these best practices can be found on the COAA Web site (www.coaa.ab.ca).

The COAA's approach to the dissemination of their best practices is to provide open and free access to all of their best practices and research results via their Web site, publications and electronic media. At their annual Best Practices Forum, the latest developments of each best practices committee are presented in the form of interactive workshops. This annual event has been held for the past 14 years and is attended by over 500 delegates.

20.4.1. Research development towards best practices

The Hole School of Construction Engineering at the University of Alberta has been conducting research for the COAA since 2000, in an effort to contribute towards the development of their industry best practices. The research needs are identified and driven by the best practices committees. The researchers serve on the committees and carry out the research in consultation with the committee members. Reporting and dissemination of the research results are done through several mechanisms: a comprehensive report to the relevant committee, presentations and workshops at the annual COAA Best Practices conference, and publications and presentations at international conferences and in refereed journals. Implementation of the research findings is carried out by construction organisations that will benefit from the relevant best practice, or the results are used towards the development of further studies. Two such research efforts are described next.

20.4.2. Effective integration of apprentices in industrial construction

Construction labour force training and effective utilisation is a significant issue facing the Canadian construction industry, particularly in the province of Alberta. A large volume of construction projects are planned over the next five years, partly driven by rising energy prices, creating a significant demand and potential shortage of skilled tradespeople. Oil sands investment alone is expected to exceed $8 billion annually from 2006 to 2014, well over the previous $6 billion peak (Construction Sector Council, 2006). This problem is compounded by the fact that the skilled labour pool is aging and nearing retirement, with the rate of labour force retirements in Alberta expecting to increase sharply after 2006 (Construction Sector Council, 2006). The number of construction craft personnel required to meet the demand for major industrial projects in Alberta is forecast to remain between 32 000 and 36 000 from 2008 to 2010, making the Alberta construction labour market the tightest it has ever been (COAA, 2006).

The province of Alberta operates on an apprenticeship system for certified trades, requiring an average of 4 years of technical and on-the-job training before becoming a certified journeyman. A journeyman is a tradesman who has been trained and certified through the province's apprenticeship and industry training system, which uses a combination of on-the-job-training, work experience and technical training. The majority of this time is spent on the job. The Alberta Apprenticeship and Industry Training Board of the Alberta Government sets minimum ratios for the number of journeymen per apprentice that are allowed on site for different trades. These ratios have recently been revised to allow for greater utilisation of apprentices. Traditionally, the industrial construction sector has had a lower rate of apprentice utilisation than other sectors of construction, such as the commercial and institutional building sectors, due to a number of barriers. The main concern in the industrial sector is ensuring the safety of often inexperienced apprentices in the highly technical nature of the work. Other barriers are related to the perception that apprentices lead to reduced cost effectiveness, due to a lack of skill, and therefore a lower productivity.

With the increasing shortage of skilled workers in Alberta, a need was identified by the Workforce Development Coordinating Committee of the COAA to develop an industry best practice on how to improve the on-the-job portion of apprenticeship training and identify means by which the industry can more effectively use apprentices. Consequently, a study was commissioned with the Hole School of Construction Engineering to develop a methodology by which to quantify and assess the impacts and benefits of apprentice utilisation on industrial construction labour productivity, with the aim of dispelling some of the perceived barriers to apprentice utilisation (Fayek *et al.*, 2003). The data collection to test the methodology was conducted in 2001 on the Athabasca Oil Sands Downstream Project in Alberta, which consisted of a 1 50 000 barrel per day bitumen upgrader. The data collection methodology and details of the results are described in Fayek *et al.* (2003).

Quantitative and qualitative data on productivity and work sampling data were collected for two trades, pipefitters and electricians, since they are two of the most significant trades in industrial construction. Four crews from each trade were studied, each for a period of three weeks. The crews chosen had different ratios of apprentices to journeymen and were conducting similar tasks at the same time under similar conditions (e.g. same time and same area on the project). This methodology minimised the variation in productivity between crews due to weather and site conditions, and also ensured a greater variety of tasks were observed. By examining crews with different ratios, the impact on productivity of the ratio of apprentices to journeymen performing a task was assessed.

The productivity and cost data showed that apprentices can be as productive and cost-effective as journeymen at specific tasks, provided they are given adequate instruction and supervision and that they are aware of proper safety practices. A sample of such data is shown in Figure 20.3 and Figure 20.4. Figure 20.3 shows the productivity of electricians associated with installing cable tray, using different ratios of apprentices to journeymen. Figure 20.4 shows the associated unit labour costs for the corresponding tasks. As illustrated in these figures, in some cases using apprentices led to equal or better productivity and improved cost-effectiveness. Similar results were obtained for the pipefitter tasks. With sufficient data collected over time, such an analysis can be used to determine the break-even ratio of apprentices to journeymen for maximum productivity and cost-effectiveness. It can also be used to develop industry benchmarks for the productivity and unit labour cost of individual construction tasks. The work sampling results showed that both electrician and pipefitter apprentices were involved in comparable proportions of direct (i.e. productive) work as were journeymen in those trades.

In addition to replenishing the workforce, utilising apprentices on industrial construction projects provides them with experience in the full range of tasks they would be expected to undertake as journeymen. Two of the recommendations arising from this study (Fayek *et al.*, 2003) are as follows: (1) to implement a planned and managed rotation of apprentices through key work assignments and types of projects, and (2) to include structured on-site mentoring programs to ensure the proper and safe transfer of skills to apprentices. The findings and recommendations of this study are now being used by the COAA and by government organisations in the development of best practices for effective apprentice utilisation and career development.

20.4.3. Measuring and classifying construction field rework

The oil sands projects in Alberta are classified as 'mega projects'. Mega projects can be defined as 'major ... projects that cost more than $1 billion, or projects of a significant cost that attract a high level of public attention or political interest because of substantial direct and indirect impacts on the community, environment and State [provincial] budgets' (Capka, 2004, http://www.tfhrc.gov/pubrds/04jul/01.htm). While mega

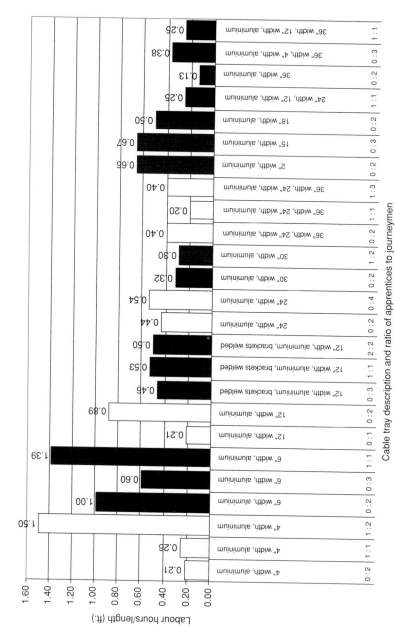

Figure 20.3 Productivity of install cable tray tasks by ratio of apprentices to journeymen.

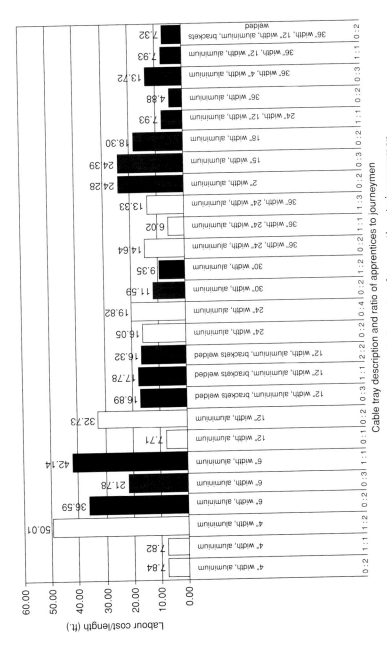

Figure 20.4   Unit labour cost of install cable tray tasks by ratio of apprentices to journeymen.

projects create significant economic and employment opportunities, they present a number of challenges, including a tendency to be prone to cost overruns. Research has shown that cost overruns on mega projects are the norm, and typically range from 40% to 200%, regardless of the geographical location of the project (Morris and Hough, 1991). Oil sands mega projects in Alberta are no exception to this trend. Cost overruns on recent mega projects have led the COAA to undertake a number of initiatives to help improve both the cost performance and predictability of mega projects.

One factor that often contributes significantly to cost overruns is construction field rework. A direct correlation between cost and schedule growth and rework has been determined (Love, 2002). Despite the significance of rework, there are few industry standards available for defining, quantifying and classifying field rework. As a result, the COAA Field Rework Committee commissioned the Hole School of Construction Engineering to undertake a study to develop the following industry standards related to field rework: (1) a definition of construction field rework (i.e. what is included and what is not), (2) an index for quantifying construction field rework, and (3) a classification system for identifying the causes of field rework. The detailed study and its findings are found in Fayek *et al.* (2004).

On the basis of this study (Fayek *et al.*, 2004, p. 1079), field rework is defined as: 'activities in the field that have to be done more than once in the field, or activities which remove work previously installed as part of the project regardless of source, where no change order has been issued and no change of scope has been identified by the owner'. Furthermore, field rework does not include: (1) project scope changes, (2) design changes or errors that do not affect field construction activities, (3) additional or missing scope or (4) off-site fabrication or off-site modular fabrication errors that do not affect direct field activities. A construction field rework index (CFRI) was proposed as a standard method of measuring field rework, as defined in the equation below:

$$CFRI = \frac{\text{Total direct plus indirect cost of rework performed in the field}}{\text{Total field construction phase cost}}$$

The classification system used for categorising the causes of field rework was based on the fishbone (i.e. cause and effect) classification system developed by the COAA Field Rework Committee, as shown in Figure 20.5. A comprehensive list of third level (i.e. root) causes was developed to attribute to rework incidents. In cases where several root causes led to a rework incident, a standard approach, based on the analytic hierarchy process (Saaty, 1980), was proposed for attributing and apportioning multiple root causes to a rework incident. A standard approach for collecting and analysing field rework data was developed and implemented in the field rework data collection system (FRDCS), the main screen of which is shown in Figure 20.6.

The field rework data collection methodology was validated in 2002 on the Syncrude Aurora 2 oil sands project in Alberta, which consisted of a mining expansion to process 58 million tons/year of ore to provide 38 million barrels/year of feedstock for a related upgrader expansion project. The methodology, definitions and classification system developed were found to be very effective. The methodology developed can also be extended to address the issue of engineering design rework. Several EPC firms are now using the FRDCS to track rework on their own projects. By making these aspects of rework measurement and classification standard within the industry and collecting

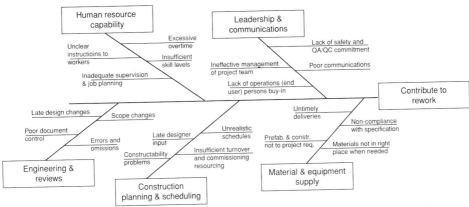

Figure 20.5 Rework cause classification system used in fieldwork study. (Source: Figure 1 ofFayek *et al.*, 2004.)

data over time, we can identify field rework levels and trends within the industry and formulate strategies to deal with the most significant causes of field rework.

20.4.4. Best practices implementation challenges

While the approach taken by the COAA for developing and disseminating their best practices has been very effective, implementation of best practices on a widespread

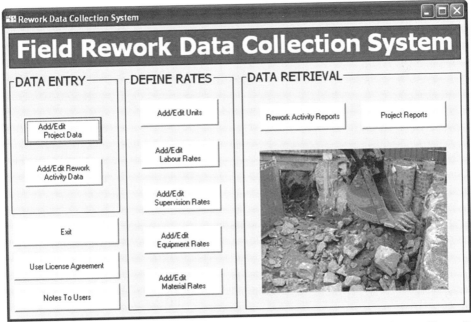

Figure 20.6 Main screen of field rework data collection system (FRDCS). (Source: Figure 3 of Fayek *et al.*, 2004.)

basis remains a challenge. One of the barriers to implementation is the fact that many of the best practices developed, although owner driven, are applicable to contractors' and EPC firms' practices. The COAA is in the process of canvassing its members to determine the barriers to effective implementation of best practices. Preliminary findings indicate that one way to increase the widespread use of best practices is to deliver in-depth hands-on training workshops of the tools developed, and to illustrate the potential financial benefits of implementing such best practices. Owners may also examine possible incentives to encourage contractors and EPC firms to implement best practices on their projects; these incentives could be in the form of mandatory requirements of contractual agreements, financial incentives or preferential awarding of contracts.

20.5. Conclusions

This chapter has presented two client driven initiatives aimed at introducing innovation to the Canadian construction industry. The CCIC, with the support of public sector clients, has undertaken a pilot study to help determine appropriate metrics, a data collection methodology and data analysis techniques to measure and benchmark the performance of the Canadian construction industry. The COAA has a long history of initiating, developing and implementing construction industry best practices to help deliver innovation to the construction industry. Both of these organisations aim to develop construction industry wide standards for performance improvement. Standardisation of any performance measure is critical to the ability to repeat, benchmark and compare the measure between projects, organisations or industries. This chapter has also illustrated how industry expertise can be combined with university-based research to deliver innovation to the construction industry. The approaches presented in this chapter provide a model of how clients can influence and drive innovation within the construction industry.

References

Capka, J.R. (2004) Megaprojects – they are a different breed. Federal Highway Administration (FHWA), *Public Roads* **68**/1, 1–15. Available at http://www.tfhrc.gov/pubrds/04jul/01.htm.

Construction Owners Association of Alberta (COAA) (2006) *Alberta Industrial Construction Projects >100 MM Cdn 2004–2010.* Available at http://www.coaa.ab.ca/LinkClick.aspx?link=pdfs/AIC_Projects_Graph-Oct-2006.PDF&tabid=81 [Accessed 3 November 2006].

Construction Sector Council (2006) *Construction Looking Forward: Labour Requirements From 2006 to 2014 for Alberta.* Ottawa, Ontario, Construction Sector Council, pp. 1–28.

Fayek, A.R., Dissanayake, M. and Campero, O. (2004) Developing a standard methodology for measuring and classifying construction field rework. *Canadian Journal of Civil Engineering* **31** (6): 1077–1089.

Fayek, A.R., Shaheen, A. and Oduba, A. (2003) Results of a pilot study to examine the effective integration of apprentices into the industrial construction sector. *Canadian Journal of Civil Engineering* **30** (2): 391–405.

Love, P.E.D. (2002) Influence of project type and procurement method on rework costs in building construction projects. *Journal of Construction Engineering and Management* **128** (1): 18–29.

Manseau, A. and Seaden, G. (eds) (2001) *Innovation in Construction: An International Review of Public Policies.* New York, NY, Spon Press.

Morris, P.W.G. and Hough, G.H. (1991) *The Anatomy of Major Projects: A Study of the Reality of Project Management.* Chichester, John Wiley & Sons.

Rankin, J. and Luther, R. (2006) The innovation process: information and communication technology (ICT) adoption for the construction industry. *Canadian Journal of Civil Engineering* **33** (12): 1538–1546. [Special issue on Construction Engineering]

Rogers, E. (1995) *Diffusion of Innovations*, 5th edn. New York, NY, Free Press.

Saaty, T.L. (1980) *The Analytic Hierarchy Process: Planning, Priority Setting, Resource Allocation.* New York, NY, McGraw-Hill.

Tooles, T.M. (1998) Uncertainty and home builders' adoption of technological innovation. *Journal of Construction Engineering and Management* **124** (4): 323–332.

Tornatzky, L. and Fleischer, M. (1990) *The Process of Technological Innovation.* Toronto, Lexington Books.

Webster's New Millennium™ *Dictionary of English, Preview Edition (v 0.9.6)* (2006) Lexico Publishing Group, LLC. Dictionary.com. Available at http://dictionary.reference.com/browse/best practice [Accessed 3 November 2006].

21 Public policy, clients and the construction industry

Eileen Fairhurst

21.1. Introduction

This chapter offers a case study of a major capital development programme, the Manchester, Salford and Trafford Local Improvement Finance Trust (MaSTLIFT), intended to address inadequate primary care estate and improve access to health and social care in the UK. The MaSTLIFT involves public–private partnership working between the UK National Health Service (NHS), the construction industry and local government. It is, then, part of that growing trend of public–private procurement as an important feature of public policy. The innovative feature of local improvement finance trusts (LIFTs) is that they are a procurement mechanism intended to improve primary care estate in under-doctored areas. Unlike private finance initiative schemes, ownership of assets is not transferred exclusively to the private sector.

Partnerships for Health, the company established to deliver the LIFT initiative, note that as assets are not exclusively owned by the private sector, the public sector are able to concentrate on service strategy and commissioning. They continue (Partnerships for Health, undated, p. 4):

> The strategic partnership has overcome the historic un-coordinated nature of service delivery in community health and social care. It establishes a Strategic Partnering Board (SPB), which includes all local stakeholders involved in health and social care. The SPB is a unique forum which is set up to collate the service requirements of all health and social care providers in the LIFT area and to plan and deliver an integrated services strategy.

This declaration, though, assumes that the public sector is a homogeneous category. In practice this may not be so. MaSTLIFT was amongst the first wave of LIFTs, which were established in 2001. It is the largest LIFT programme in the country. Originally, there were ten public partners: Manchester City Council, Salford City Council and Trafford Metropolitan Borough Council (MBC) from local government and North, Central and South Manchester Primary Care Trusts (PCTs), Salford Teaching PCT (Salford tPCT), North and South Trafford PCTs and Greater Manchester Ambulance Service (GMAS) from the NHS. With the national re-configuration of PCTs from October 2006, the number of PCTs has reduced from six to three (Salford, Manchester and Trafford) and, consequently, the number of public partners from ten to six. (GMAS ceased to exist with the establishment of the North West Ambulance Trust). In July 2004, the LIFT Company (LIFTCo) legally established the public–private sector partnership. MaST LIFTCo is a property development joint venture company with public sector partners, Partnerships for Health and Excellcare (a joint venture between Equion, a division of John Laing PLC and Bank of Scotland), as shareholders.

216

The number of public partners involved in MaSTLIFT adds to the complexity of partnership relationships not only across the public–private sectors but also within the public sector itself. This points to the problematic nature of the category of 'client'. Whilst there is a growing literature on market relationships in public sector procurement (Allen, 2002; Erridge and Greer, 2002; Turner and Martin, 2004; Bovaird, 2006), there is little evidence specifically on LIFT companies of which there are currently 42. The NHS Service and Delivery (SDO) R&D Programme issued a call in 2006 for research proposals on LIFT schemes but the evidence from studies is awaited. This chapter, written from the perspective of a participant in the activities of the MaSTLIFT, is offered as a study in organisational development and adds to the growing literature on social science perspectives on the construction industry (Bresnen *et al.*, 2005a, 2005b; Rooke *et al.*, 2005). In terms of Bovaird's (2006) analysis, the focus here is on partnership procurement. The substantive matters to be examined here relate primarily to Salford tPCT in the context of MaSTLIFT. This is not just an idiosyncratic matter as Salford tPCT, of all the public sector partners, arguably had a greater interest in the realisation of LIFT. On the contrary, following this path allows the implications of multi-stakeholder public partners for the wider public–private partnership to be identified, demonstrates the fluidity of relationships involved and the emergent/context bound features of the category 'client'.

The chapter has three major sections. Initially, the provision of health and social care services in Salford are located within the context of national public policy in which focus is on health inequalities, partnership working and public involvement in decision-making. Attention is then directed towards a socio-demographic portrait of Salford and the challenges presented to health and social care provision. Such an approach links to the chapter's second section where the term 'client' is explored. Whilst LIFT was established as a procurement mechanism, its translation into tangible outcomes is contingent upon effective relationships between client and supplier. The ways in which Salford tPCT engaged with different kinds of 'clients' will be outlined. The chapter concludes by drawing out transferable learning to other public–private partnerships in which the construction industry may be involved.

21.2. Public policy and health in Salford

A hallmark of the current government's public policy has been the modernisation of public services; no more so than in the realm of health policy. The publication of *The NHS Plan* in 2001 contained a 10 year plan to modernise health services (Department of Health, 2001). It was a policy response to the 'post code lottery' whereby access to services depended upon place of residence and the consequent variation in access to those services. Such inequitable access compounded the correlation between place of birth and life expectancy. In addition to the introduction of targets for reducing waiting lists to improve access to secondary care, attention was directed, with the implementation of National Service Frameworks and the establishment of the Healthcare Commission, to standards for, and quality of, services. One of the ways in which variations in access to primary care was addressed was via the establishment of the LIFT mechanism. All of these measures aimed to reduce health inequalities. At the same time there has been increased expenditure on health: total UK healthcare spending will have increased from £65.4 billion in 2002–2003 to £105.6 billion in 2007–2008.

The intention of reducing health inequalities was emphasised further with the Wanless reports (2002, 2004) on public health. Wanless (2004) argued that the costs of the NHS were not sustainable and that it was the NHS's responsibility to secure long term health gain. Discharging the latter entailed addressing the determinants of health, for example, housing, education and employment, or 'upstream' interventions. Achieving long term health improvement was contingent upon not only these 'upstream' interventions but also individuals becoming 'fully engaged' in their health. An example of this is enabling people with long term conditions to manage, through schemes such as the 'expert patient' programme, their own illness. The more recent white paper 'Our Health, Our Care, Our Say' (Department of Health, 2006) identified patient and public involvement as a necessary feature of commissioning decisions to be taken by PCTs.

This current public policy provides the backdrop against which the health of Salford's population may be placed. Salford is ranked, using the Index of Multiple Deprivation 2004, as the twelfth most deprived local authority in England although there is a small number of areas, which are amongst the most affluent in the country. A consequence of this high level of deprivation is that life expectancy is lower in Salford than the average for England and Wales. For instance, for births between 2001 and 2003, Salford men can expect to live for 2.9 years less than the average for England and Wales. Premature mortality in Salford is accounted for by above average deaths as indicated by the standardised mortality ratios for cancer and circulatory diseases. Salford has higher rates of ill health, as indicated by long term limiting illness reported in the 2001 Census, compared with the rest of England and Wales (Salford PCT, 2006). Given the health challenges evident in Salford's population and that there was a history of under investment in primary care how did Salford engage in the MaSTLIFT scheme?

21.3. 'Clients' and partnership working

Bovaird (2006) identifies multi-stakeholder agreement of strategic priorities as a particularly problematic aspect of partnership procurement. His case study demonstration of this involves one NHS organisation and one local authority. In the case of MaSTLIFT, though, the category of 'client' is more complex and its unpacking adds to our understanding of an ostensibly bi-lateral client–supplier relationship. There are, analytically, at least three kinds of client relationships apparent in MaST LIFTCo: firstly, within each local authority's geographical area between the local authority (LA) and the NHS; secondly, between Manchester's, Salford's and Trafford's LAs and PCTs at the Strategic Partnering Board (SPB); and thirdly, through the public sector 'shareholder', with the LIFTCo Figure 21.1 represents these different kinds of client relationships.

21.3.1. NHS and local authority as 'client'

When established in April 2001, Salford PCT was keen to become the leaders in the health economy by embracing their role as a public health organisation, responsible for improving the health of the population. The PCT was able to build upon two existing initiatives in the city: Charlestown and Kersal New Deal for Communities (NDC), a government sponsored regeneration programme, and Salford Health Investment for Tomorrow (SHIFT). These two initiatives were used as opportunities to make

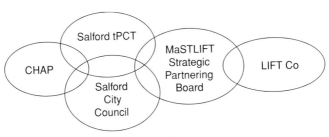

Figure 21.1   'Client' Relationships with LIFTCo

connections with LIFT and, thereby, make it a reality. Although the former initiative has come to fruition, that with SHIFT is much more ambitious and is currently in the midst of development.

The Charlestown and Kersal NDC had supported the establishment of a voluntary group called Community Health Action Partnership (CHAP). This arose from local residents who were concerned about poor access to general practitioners in their area and wanted to be involved in the planning of new services. CHAP's concerns reflected those of the PCT and opportunities presented by LIFT were used to meet the needs of Salford residents. In effect, the PCT were able to 'piggy back' on the NDC, which had an explicit focus on regeneration. Two healthy living centres are now operating in the NDC area. The development of these two centres demonstrates how partnership working may result in 'collaborative advantage' (Huxham, 1995) so that improvements in health may be co-produced by professionals and lay people. In these cases, residents of the area gained improved access to primary care, an 'expert patient' programme run by local residents, exercise classes run by the leisure services of the council and community rooms. These examples of partnership working resulted in services being delivered in different ways and in different places.

Working with the NDC resulted in an 'early win' but building upon SHIFT is much more complex. In February 2001, prior to the establishment of Salford PCT, the Strategic Outline Case for SHIFT was approved. This proposed re-development of an acute hospital together with 'primary care resource centres' in which hospital services could be more appropriately accommodated. Arguably, primary care was an adjunct to the SHIFT programme. Between 2001 and March 2003 when the Outline Business Case (OBC) for SHIFT was approved by the former Greater Manchester Strategic Health Authority, the original proposals were turned round from being primarily about hospital re-provision and, specifically, NHS driven, to whole system reform including the LA. The 'primary care resource centres' became health and social care centres. In this way, LIFT was transformed from being solely a procurement mechanism for new primary care estate to being a vehicle through which public sector services could be re-designed. SHIFT and LIFT programmes became interlinked so that services traditionally found in hospitals, for example, rehabilitation and diagnostics will be re-located in LIFT centres. The scale of this partnership between an acute hospital and a PCT is unique in the North West.

The vision for the health and social care centres grew out of interactions between the NHS and the LA. In that sense, the vision emerged over time. The first step was deciding: how many LIFT centres to build and where to locate them; the city council played an important role in this. In many ways, identifying the number of centres was serendipitous. The city council were developing four distinctive 'townships', based upon the previously separate local government areas that had enlarged the City of Salford in 1974. Shops, public services and public transport were at the core of each. They owned land, which could be developed. These four townships provided ideal locations as they had a ready-made footfall, which would engender access to services.

Once the locations had been identified, a joint PCT/City Council programme of public and staff engagement was started. Initially this followed the conventional public meeting format but was then tailored to specific topics and interest groups. A database, compiled from attendees, was established and contacts were periodically made with them. Information gathered from public engagement activities has fed into decisions about which services are to be located in particular centres. Each of the four centres is to have a specialist focus in terms of the provision of services, which currently are located in the acute trust. Other NHS services to be housed in them will be general practice, dentistry, pharmacy and community nursing and allied health professionals. As in the two centres already operating, there will be community rooms and cafes. The innovative features of the centres are proposals for integrated working, as opposed to mere co-location, between public sector partners. City council one stop shops and libraries also will be in some of the centres but, rather than each organisation's services having their own separate reception desks, there will be one for the whole of the centre. Financial closure has been achieved on three of these four centres and preliminary site work has started.

The transformation of the original SHIFT scheme into whole system design and the inclusion of a community group (CHAP) into the MaSTLIFT were enabled not only by the changing policy context, outlined earlier, with its more explicit focus on health improvement but also because of the high level of commitment and willingness to innovate of senior leaders in the NHS and LA. For instance, the re-provision of some library services to the new centres was not 'in the frame' at the outset but resulted from a 'wouldn't it be a good idea' perspective. A consequence of this, too, was that city council involvement was at a corporate level rather than restricted to a more conventional relationship with social services. The city council's commercial expertise and experience, gained from their previous involvement with the private sector, was much needed in the NHS where there was little. Overall, MaSTLIFT became part of the wider regeneration of Salford.

Arguably, innovation may be contingent upon risk taking and was evident in the actions of leaders in the PCT. Firstly, the Chief Executive of the PCT had been involved in the development of the LIFT initiative at central government and, consequently, identified its potential for changing delivery of services. Secondly, whilst PCTs were new and immature organisations, both the Chief Executive and Chairman of Salford tPCT were very experienced NHS leaders. Consequently, advantage was taken of the NHS bureaucracy and, rather than continually 'seeking permission' to pursue particular courses of action, action was taken.

21.3.2. MaSTLIFT public sector partners as 'client'

The extent to which LIFT has been embraced by the Salford public sector has not been matched yet by the other public sector partners. One indicator of this is the capital value of schemes: Salford's tranche 1 schemes are worth £50.85M. This represents 61%

of the total capital value (£83.90M) for the MaST schemes. Given that the achievement of many of Salford's wider objectives are locked into the completion of LIFT schemes, it is in Salford's interests that MaSTLIFT is effective. This has encouraged Salford PCT to act as a 'project champion' (Noble and Jones, 2006).

The SPB occupies a pivotal place in the LIFT model of procurement for it is responsible for developing the client strategy to which the supplier responds. The involvement of all public sector partners in the SPB was variable so that high-level attendance at meetings was not consistent. This lack of coherence meant that the SPB needed to learn to be a 'client'. This points to the important issue of governance. A development day was held at which the roles and responsibilities of the SPB were clarified. Salford tPCT Chair convened regular meetings with her Chair colleagues to engender high-level organisational commitment to making the SPB work. The 'jelling' of the SPB into an effective 'client' has been hampered by events in the external environment. The NHS and PCTs, in particular, have been subject to yet another re-structuring. Six of the original NHS partners no longer exist. Such matters introduce turbulence into organisations as individual's interests are diverted to the consequences of impending change.

21.3.3. Public sector 'client' and the LIFTCO

Just as stability of personnel amongst public sector partners may contribute to their actions as a 'client', so is this matter relevant to the private sector. A number of changes of private sector personnel resulted in relationships between them and public sector partners having to be established anew. This has inevitable consequences for trust, a cornerstone of effective partnership relationships (Fairhurst, 2003). Although partners may use the same language of partnership, the meanings of those words may not necessarily be the same for meanings are situated in social contexts. As Bovaird (2006) demonstrates, partnership procurement is more than a technical matter; it is socially constructed from the kinds of interactions and negotiations outlined in this chapter. Moreover, the conventional 'rules of the market' and associated behaviours in, and with, which the private sector are used to operating may be inappropriate to partnership procurement and may require 'loosening up'.

21.4. Transferable learning

Now that different kinds of 'client' relationships, salient to the realm of partnership procurement have been outlined, it is possible to suggest some transferable learning arising from this exploration. In relation to public sector partners, the UK NHS has much to learn from the experience of their LA colleagues about working with the private sector. Public sector partners may not share the same interests and the implications of this for their contribution to making partnering 'happen' requires acknowledgement. Learning to be a 'client' is the outcome of a shared development process and high-level organisational commitment. Just as being 'a client' is a contextual matter within temporal parameters, so is being a 'project champion'. Where multi-stakeholders constitute public partnerships, different project champions are likely to emerge at different times. This, in turn, has implications for relationship building and interactions between the public and private sectors; rather than being a once-and-for-all issue, client–supplier relationships require continual 'work'.

For private sector partners, the recognition that the practices of public sector procurement are a social production from interactions with a range of stakeholders, has consequences for 'rules of the market'. In the case of partnership procurement, market rules are not given; on the contrary, political dynamics of the situation are crucial for their understanding. In this sense, too, clients driving innovation is a continuous process. These distinctions between procurement as a technical and socially constructed matter may account for claims made by both client and supplier of each other as 'lacking in trust'. Finally, despite some features of client–supplier relations resulting in 'challenging' relationships, all the first tranche of MaSTLIFT have reached financial closure and centres currently operating were built on time and within budget.

References

Allen, E. (2002) *Managing Strategic Service Delivery Partnerships from Governance to Delivery.* London, New Local Government Network.

Bovaird, T. (2006) Developing new forms of partnership with the 'market' in the procurement of public services. *Public Administration* **84** (1): 81–102.

Bresnen, M., Goussevskaia, A. and Swan, J. (2005a) Managing projects as complex social settings. *Building Research Information* **33** (6): 487–489.

Bresnen, M., Goussevskaia, A. and Swan, J. (2005b) Implementing change in construction project organisations: exploring the interplay between structure and agency. *Building Research and Information* **33** (6): 547–560.

Department of Health (2001) *The NHS Plan.* London, HMSO.

Department of Health (2006) *Our Health, Our Care, Our Say.* London, HMSO.

Erridge, A. and Greer, J. (2002) Partnerships and public procurement: building social capital through supply relations. *Public Administration* **80** (3): 503–522.

Fairhurst, E. (2003) Context, typificatory processes and health and social care partnerships. In: *Collaboration in Context* (eds C. Scott and W. Thurston). Calgary, Alberta, Health Promotion Research Group and Institute for Gender Research, University of Calgary., pp. 94–101.

Huxham, C. (ed) (1995) *Creating Collaborative Advantage.* London, Sage.

Noble, G. and Jones, R. (2006) The role of boundary-spanning managers in the establishment of public–private partnerships. *Public Administration* **84** (4): 891–917.

Rooke, J., Seymour, D. and Fellows, R. (2005) Learning, knowledge and authority on site: a case study of safety practice. Building Research and Information **33** (6): 561–570.

Salford Primary Care Trust (2006) *The Changing Picture of Health in Salford: Public Health Annual Report 2004–2005.* Salford, Salford Primary Care Trust.

Turner, D.N. and Martin, S. (2004) Managerialism meets community development: contracting for social outcomes. *Policy and Politics* **32** (1): 21–32.

Wanless, D. (2002) *Securing Our Health: Taking a Long Term View.* London, The Treasury.

Wanless, D. (2004), Securing good health for the whole population, The Treasury, London.

22 Value for money versus complexity: a battle of giants in the public sector?

Erica Dyson

22.1. Accountability and value for money

Everyone has heard the story of the man who was down on his knees in the middle of a dark street, when someone came along and asked him what he was doing. 'I've lost my keys and I'm looking for them'. 'Well, are you sure you lost them around here?' asked the other person. 'No, I'm not at all sure' said the man, 'But if I don't look for them here whilst there's some light, I won't have a cat in hell's chance of finding them anywhere else'.

A stab in the dark, is a stab in the dark, no matter how we rationalise it. Not all effort is productive, no matter how well intentioned. When clients driving innovation are determining the business benefits of a given investment, they need to be sure that they are not just telling a good story. Further, management accountability demands that business cases and value for money statements be seen as robust and reliable, not stabs in the dark or honourable guesses. In other words, it is critical that public as well as private sector decision makers see themselves as agents for driving innovation. Conversely and as importantly, account managers providing services or products to the public sector should view those same individuals as clients for whom a key part of their role is to drive forward innovation in partnership with them.

From an investor perspective, robust business cases are also critical. For investors, whether they are shareholders or tax payers, need value for money. It is axiomatic: no one wants to pay more than is necessary and whatever the level of investment, we want to be assured that the outcome anticipated will be achieved and that investments will stand up to rigorous value for money tests both ante and post hoc.

But how can clients driving innovation be assured of value for money when, in the public sector arena, they are investing in *new* or *different* ways of achieving outcomes? They invest in new and different ways because they believe that more public good can be delivered through these new methods. This gives rise to another problem related to the likelihood of outcomes being as intended. When the combination of inputs is different to previously, managers as clients *cannot be sure of an outcome*. This is because of the laws of complexity, which guarantee an uncertainty of result together with the certainty that there will be unintended outcomes. By definition, they cannot know what the unintended outcomes will be. Therefore, it is not possible to put a value on the extent to which unintended consequences will have a positive or negative impact upon the original purpose of a said investment.

Where a principal driver for public sector investment is improving the public good, our investments must involve change. Therefore, complexity, uncertainty and risk cannot be avoided. They are part of the investment and change process. Value for money tests must, therefore, take this into account and factor in complexity as a given.

This then gives rise to a governance conundrum. The problem is that if uncertainty of outcome is unavoidable, and managers cannot know what the outcome will be, is it right that they should be held to account over the value for money of a given investment? Furthermore, if clients or investors cannot be sure of the outcome, is it, therefore, impossible to engage in change, which by definition is complex and uncertain, and achieve value for money at the same time? Is this a forced choice where there is either complexity in schemes which it is hoped will deliver more public good or there are value for money investments which are based on 'more of the same'. In other words, does it make any practical sense to hold anyone accountable? For, if the public cannot hold public sector managers to account, who understand the change proposals in some detail, what value is there in holding the manager's manager to account when he/she knows much less about the specifics? In fact, can anyone ever be held to account for the investments relating to improving for the public good? In what sense is the concept of accountability valid or is it another good story to soothe taxpayers' nerves?

Instinctively, the answer to the last question must be that it is entirely appropriate to hold decision makers to account on the grounds of value for money being achieved or not. However, is this just a moral view? To what extent can value for money (VFM) assessments be based on firm foundations so that holding to account is not just a 'blood letting process' but one that is valued by managers and the public because it leads to better outcomes over the longer term? Empirically, the tension between achieving consistent VFM and delivering complex change lies in another dimension, more to do with organisational behaviours and culture than the complexity of the change.

Using examples from experiences in a large-scale local improvement finance trust (LIFT) partnership, this chapter proposes an approach to increasing the likelihood of securing VFM. It does this by linking strategic objectives to VFM performance outcomes along with an assessment of the organisational maturity and capability of those delivering the solutions.

22.2. The link between capability and outcome

Do we know a capable person when we meet one? How many capable people does it take to make a capable organisation or a capable team? Or conversely, how many people does it take to negate the good work of others? There is no simple answer, of course.

The capability maturity model (CMM), first developed for the software industry, has become a useful starting point. The CMM provides organisations with a framework for assessing its systems and processes for achieving quality and for assessing the extent to which the processes it employs are both efficient and effective. Five levels are applicable, level one being the lowest level of maturity and competence.

At maturity level 1, processes are usually ad hoc and the organization usually does not provide a stable environment. Success in these organisations depends on the competence and heroics of the people in the organization and not on the use of proven processes. In spite of this ad hoc, chaotic environment, maturity level 1 organisations often produce products and services that work; however, they frequently exceed the budget and schedule of their projects. Maturity level 1 organisations are characterised by a tendency to over commit, abandon processes in the time of crisis, and not be able to repeat their past successes again. Conversely, level 5 organisations, of which there are very few, are characterised differently. They have well established evidenced based change processes for delivering continuous improvement at all levels of the organisation. This enables these organisations to continually revise strategy and business practices to reflect changing business objectives. Both the defined processes and the organisation's set of standard processes are targets of measurable improvement activities. A core characteristic of level 5 organisations is an empowered workforce where decisions are taken at the lowest level appropriate to the risk.

The model has been adapted by the University of Salford, UK, in the SPICE 3 research project[1], which looked at refining processes at the project level as well as at the organisational level. The CMM sets out various attributes and characteristics of the software development process providing a basis for assessing the extent to which information technology solutions were likely to be reliable, robust and to deliver to budget. The model has now been used by many other industries. In fact, Kelly and Allison (1999) have devoted a whole book to adapting CMM to gain a greater understanding of organisations in general. All the adaptations share the common approach of focusing on not only on business processes but also, and crucially, on *business behaviours*.

Process modelling and process improvement initiatives have shown that there is an inextricable link between outcomes, processes and people. This is true even where processes are highly automated such as in manufacturing sectors. In the case of health, social and economic regeneration, which is the purpose of LIFT public–private partnerships, very little is automated. Action research undertaken by the University of Salford into the requirements capture process, for example, supports this finding (Tzortzopoulos *et al.*, 2005). Thus, the *dominant* factor affecting the extent to which processes are effective is the behaviour of individuals and teams involved in solution delivery. Critically, behaviour affects the extent to which VFM is being secured.

Repeatability is a key component of sustainability. Thus, the ability of an organisation to be capable of learning from its own work as well as those of others is a crucial factor in securing VFM. Where an organisation lacks the ability to self-correct, any good that is achieved on its projects will not be repeated on future schemes, unless by chance. Thus, *organisational capability* is critical to gauging the likelihood of business benefits being secured. This too has been the experience of new ventures, such as the LIFT public–private partnerships, and there is emerging evidence that the partnerships of these types, which are building their capacity for learning and self-correction, are delivering more schemes to time and budget.

In their book 'The Competitive Advantage', Kelly and Allison (1999) have set out a model for matching behaviours to the respective level of organisational maturity. From the experience of the Manchester Salford and Trafford (MaST) LIFT public–private partnership, a level of sophistication has been observed in that organisations can be at one level of maturity for process A and at another level for process B. This does not deflect or detract from the overall validity. Indeed, it can enhance and enrich understanding of our collective strengths and weaknesses. It may further pinpoint areas of success or improvement to particular teams of people or departments rather than ascribing a maturity level to the organisation as a whole. Table 22.1 sets out the five levels of organisational maturity taken from the CMM.

Table 22.1   CMM capability maturity levels

Level	Capability maturity	Spread	Measurement
5	Consciously competent autopoiesis[a]	Enterprise	Tracks patterns in enterprise and co-evolution with environment
4	Quantitatively guided self–organisation	Enterprise	Models and analysis based on statistical processes
3	Guided self–organisation	Unit	Tracks team performance against intentions at multiple levels; links micro and macro emergence
2	Conscious self–organisation	Team	Gauges team capability as performed by game plan
1	Unconscious self–organization	Individual	Haphazard data about unknown behaviour patterns

Source: Kelly and Allison (1999). Reproduced with kind permission of McGraw Hill.
[a]From the Greek meaning the capacity to self-generate.

Kelly and Allison (1999) then go on to set out various behaviours pertaining to individuals (agents), teams, business units or enterprises at each of the five levels of capability maturity. By matching those behaviours to the delivery or host organisations engaged in a given investment, a more realistic assessment of benefits can be determined. Table 22.2 shows examples of behaviours for organisations with the lowest capability maturity level, level 1. Table 22.3 sets out the behaviours for organisations that are working effectively and have consistent learning loops, labelled 'guided self-organisations'. Table 22.4 describes behaviours for the top level of maturity.

Once the level of capability has been identified, then benefits can be 'matched' to that level. This is the approach that is being used 'in reverse' at MaST LIFT. A partnership-wide balanced scorecard has been generated as a tool to assist the partnership in gauging the extent to which it is achieving VFM.

In common with all balanced scorecards, it has four dimensions seeking to address four questions, which are set out graphically in Figure 22.1:

- To succeed financially, how should we appear to our shareholders?
- To achieve our vision, how should we appear to our customers?
- To satisfy our shareholders and customers what business processes must we excel at?
- To achieve our vision, how will we sustain our ability to change and improve?

Again, in common with balanced scorecards, performance indicators have been selected that best address these four dimensions. Using best practice identified by the Accounts Commission for Scotland (1998) and the Audit Commission (2000) for England, indicators have been created that have the following characteristics:

Table 22.2 Examples of behaviours at level 1 – maturity unconscious self-organization

View	Dominant behaviours
Scale	Hero driven efforts predominate
Momentum Becoming	Fear, mistrust and deception drives interactions Withholding information Protecting unique knowledge Protecting unique skills Warding off change
Belonging	Agents agree to any requests Agents intend best effort Agents deliver as possible Customer often dissatisfied
Being	Agents make reactive personal decisions Agents make self-serving trade-offs Agents pay lip service to joint planning then follow personal agendas
Self-generating behaviours	Leader tries to control agent interaction by edict and demand Agents pretend to follow orders
Emergent system	Agents feel victimised Leaders feel out of control

Source: Kelly and Allison (1999). Reproduced with kind permission of McGraw Hill.

- Relevant to the organisation's strategy
- Unambiguous in their measurement
- Cost effective to measure
- Simple to understand

However, linking performance indicators that are unambiguous and cost effective together with pan-organisational objectives[2] was insufficient. This is because the indicators are linked to *best practice outcomes*, which are the very best that can be achieved. Initially, they were not linked to the partnership's ability to deliver those benefits.

Thus, a *staged approach* to assessing overall performance has been devised. This links each of the desired performance outcomes across the four dimensions with a level of organisational maturity based on the CMM levels 1–5. By so doing, the partnership is better able to predict which outcomes it is likely to be achieving and which it is not at any given stage in the partnership's implementation of its portfolio. Table 22.5 lists the scorecard indicators for a level 1 LIFT public–private partnership and for those who consider themselves at level 2.

Accurate forecasting has a number of benefits, not least of which is the boosting of team morale when successes, however small, are achieved. At the corporate level, collective understanding is increased regarding which risks are likely to be contained and which are not. Further, there is greater transparency around expectations versus actual delivery and a better understanding of just what it takes to achieve good VFM.

Table 22.3 Examples of behaviours at level 3 – maturity guided self-organization

View	Dominant behaviours
Scale	Team-to-team relationships being strengthened
Momentum	Mutual respect, trust and honesty drive agent interactions in unit
Becoming	Agents in unit openly share qualitative information Agent knowledge based on unit history Unit experience captured Agents in unit contribute to tactical innovation Agents protect unit
Belonging	Agents negotiate deliverables Agents able to keep their word Renegotiation less frequent Customer gets deliveries as expected
Being	Agents make unit rules and rule-based decisions Agents make objective unit trade-offs Agents select plays, tailors standard game plans and share feedback on results
Self-generating behaviours	Leader influences emergent unit behaviour Agents self-reinforce desirable behaviour
Emergent system	Unit empowers itself through the relationship with others

Source: Kelly and Allison (1999). Reproduced with kind permission of McGraw Hill.

A further refinement has been the linking of performance outcomes with one or more of the partner organisations so that we can be clear about where the focus for behavioural and process change is required. Knowing which processes and which organisations/teams to focus effort on becomes critical in not contributing to the 'shot in the dark' epitomised by the story of the man searching for his keys.

To summarise, the MaST LIFT balanced scorecard has a range of performance outcomes with the following features:

- Linked to one of four levels in organisational maturity using CMM.
- Linked to pan-organisational (partnership-wide) objectives.
- A top ten 'critical' set of indicators spanning all objectives[3].
- Linked to one of the four dimensions of partnership performance.
- Sub categories pertaining to the substance of the programme. In this instance, covering areas such as assurance, estates management, benefits realisation, corporate citizenship, sound financial management and robust project management.
- Indicators attached to one or more specific organisations in the partnership.
- Clear measures and criteria for indicating performance on a red, amber, green basis.

Table 22.4 Examples of behaviours at level 5 – maturity consciously competent autopoiesis

View	Dominant behaviours
Scale	Agents identify with enterprise and understand how they help the whole fit with the environment
Momentum	Double-loop learning in place
Becoming	Agents openly share with enterprise partners Agents create intellectual capital for enterprise Agents forecast probabilities and trends of future enterprise experience Agents see part of their job as enterprise innovation Agents protect enterprise ecology
Belonging	Agents brainstorm requests Agents value ideas and commitments Agents negotiate new ideas Stakeholder needs satisfied
Being	Agents make value-based enterprise decisions Agents make trade-offs for enterprise success Stable but evolving agent-to-agent enterprise interactions
Self-generating behaviours	Leaders forecast emergent enterprise and environment Agents help to reinforce or redirect needed patterns
Emergent system	Enterprise empowers itself through thinking and autonomous agents

Source: Kelly and Allison (1999). Reproduced with kind permission of McGraw Hill.

- Reporting period per indicator.
- Data sources per indicator.
- Trend analysis graphs per indicator and per partnership objective[4].

22.3. The link between capability and value for money

If the link between capability and likelihood of outcome is critical, then it is essential that decision makers understand the organisational capability for respective change programmes as they are being appraised, developed and implemented. Thus, assessing capability is not an event. Rather, it is a process that must be revisited continuously throughout the change lifecycle. Below are possible questions that would thus be pertinent at the business case stage.

- Does the organisation have experience of knowing what success will look like if the investment is successful?

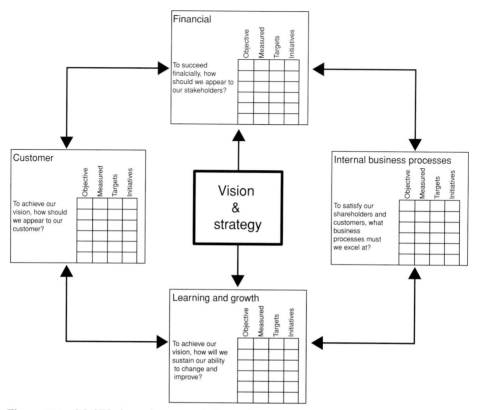

Figure 22.1   MaST balanced scorecard dimensions.

- Does the organisation have experience of knowing the errors that are likely to be made and how they could be avoided or corrected with least damage?
- Does the organisation have a track record in cost effective negotiation or problem solving that is commensurate with the proposed investment?
- How does the organisation evidence that it has learnt from successes and failures?
- Who does the organisation benchmark itself against?
- If the organisation were at a higher capability maturity level, how would the benefits be adjusted? Has the organisation risk adjusted the benefits?
- How would partnering increase the capability maturity level and, therefore, increase the likelihood of the investment benefits being secured?
- Has the organisation identified a set of conditions where it would not recommend going forward?

This approach also gives rise to additional lines of enquiry after the investment as part of the benefits realisation process:

- When were unintended outcomes identified and, whether they were adverse or favourable, how did the organisation respond?
- How have lessons learned been incorporated into new activity?

Table 22.5 MaST balanced scorecard indicators – level 1 and level 2

Scorecard dimension	Corporate aim	Performance indicator
Level 1		
Customer	Performance management	Helpdesk is accessible and communicates back to users%
Customer	Performance management	Planned preventative maintenance in period%
Customer	Performance management	Priority 1 calls completed on time
Financials	VFM	Scheme handover on time
Financials	VFM	New schemes meet external benchmarks
Financials	VFM	Shared risk management
Processes	Manage risk	SPB project assurance is fit for purpose
Processes	Manage risk	Project management disciplines/methodology active
Processes	Manage risk	Shared highlight reports are fit for purpose
Processes	Manage risk	Shared integrated programme plan fit for purpose
Level 2		
Customer	Partnering	Culture and behaviour
Customer	VFM	Partnering services–financial close management
Customer	Performance Management	Priority 2 calls completed on time
Financials	VFM	Rental within Stage 1 business case (cumulative)
Financials	VFM	Stage 2 within 1 month of target date (cumulative)
Financials	VFM	Market testing programme
Learning	Continuous improvement	Joint continuous improvement plan
Learning	Continuous improvement	Strategic partnering agreement review
Learning	Appropriate strategy	Community engagement in utilising the facility
Learning	VFM	Improved access to core services
Learning	VFM	Greater access to integrated health & local authority services
Learning	VFM	Schemes meet government and local policy drivers

- How able was the organisation in adapting to changing circumstances whether inside or outside its control?
- What would the organisation do differently if it were aiming to achieve the same outcomes again?
- At what point would the organisation deem the investment not to have been VFM?

These lines of enquiry focus on the organisation's ability to self-correct and self-improve. In no way do they substitute for assessing and evaluating whether the outcomes are indeed VFM. There will be a break-even point where the maturity-adjusted outcomes are not commensurate with the level of investment.

A further adaptation of the CMM would then be to assess the value of investing in improving organisational capability up-front in return for increasing the likelihood of achieving VFM. Were that approach adopted, it would bring added value and criticality to investing in organisational continuous improvement. Consequently, there will be corporate recognition that it is the combination of excellent organisations and excellent strategy that generate value and that neither can succeed on their own.

To summarise, innovation and investing for the public good must involve complexity with all the uncertainty and risk that this brings. The extent to which they 'fight' or neutralise one another is linked to the capacity of organisations to learn, adapt and self-correct in the face of foreseen or unintended circumstances. This means that the capacity of the organisation to achieve the desired outcome has to be under constant review. It also means that this review process should lead to the required changes in organisational structures and business culture.

The likelihood of VFM being secured for the investor is thus dependent upon a keen assessment of maturity-adjusted benefits at each stage of the change lifecycle, starting at the earliest point. This integrated approach will provide a more transparent basis on which to hold organisations to account on the extent to which investments have delivered VFM.

Lastly, here's another story. A retired archer had a fantastic reputation for repeatedly getting bulls eyes. 'What's the secret of your success?', asked a junior club member. 'Well' replied the marksman, 'I had two approaches. One was practicing really hard, getting to know the terrain, understanding my fellow competitors and their strengths and weaknesses and then shooting to the very best of my ability'. 'And the other approach?' asked the young person. 'Ah', said the retired marksman, 'well, when all else failed, I moved the target to where my shot landed. I was so well respected that no one ever suspected anything'.

Notes

[1] SPICE 3 (Structured Process Improvement for Construction Enterprises) Facilitating Organisational Process Improvement, The University of Salford, 2004.
[2] Pan-organisational because the MaST LIFT partnership comprises many organisations each with their own distinct and complex governance arrangements.
[3] The top ten indicators require a level 2 organisation using CMM.
[4] The partnership objectives are: managing risk; performance management; VFM; continuous improvement; appropriate strategy; and, effective partnering.

References

Accounts Commission for Scotland (1998) *The Measures of Success – Developing a Balanced Scorecard to Measure Performance.* Scotland, Accounts Commission for Scotland.

Audit Commission (2000) *On Target: The Practice of Performance Indicators.* England, Audit Commission.

Kelly, S. and Allison, M. (1999) *The Complexity Advantage.* New York, McGraw Hill.

Tzortzopoulos, P., Chan, P., Kagioglou, M., Cooper, R. and Dyson, E. (2005) *Interactions Between Transformations: Flow and Value at the Design Front-end for Primary Health Care Facilities.* Salford, The University of Salford.

23 The role of the professional client in leading change: a case study of Stanhope PLC

Colin Gray

23.1. Introduction

Broadgate, in the City of London, UK, was a world-class construction project, which provided standards of quality, speed of construction and efficiency rarely matched elsewhere. In addition, those involved in the project have and will continue to exert a large influence on the attitudes and performance of many parts of the British building industry. Broadgate was designed and constructed against the background of a building industry in the UK, which faced severe and justified criticism. Studies by Slough Estates (1979) and the Department of Construction Management, University of Reading (Flanagan *et al.*, 1979) had shown the UK building industry to be producing buildings of variable quality, slowly, achieving low levels of productivity and paying low wages to its operatives. The design and construction of the Broadgate project in the City of London in the 1980s was a seminal point in the evolution of the UK's modern construction industry. The project consolidated the results of a considerable evolutionary effort by a key group of people. Whilst they were principally property developers they had chosen as their main strategy to build fast and so gain a competitive edge in the market place. The success of Broadgate was considerable. It challenged all preconceptions about technology, organisation and the industry's culture. The approach to construction on Broadgate was largely determined by a team of Stanhope's Directors – Sir Stuart Lipton, Peter Rogers, Peter Kershaw, Paul Lewis and others who developed the specific details and practice of the construction management approach. Construction management is really about a management philosophy to enable the right attitude and involvement to develop. The key words used by Rosehaugh Stanhope Developments were 'use the A team', 'non confrontational attitude', 'relationship', 'seventy to eighty per cent repeat work, and safe work at all times'.

The team led by Sir Stuart Lipton and Peter Rogers have become dominant in the industry from this position of strength through performance. They have gone on to exert influence by chairing the Commission for Architecture and the Built Environment CABE (Sir Stuart Lipton) and the Strategic Forum and Constructing Excellence (Peter Rogers). This chapter examines the development of the Stanhope philosophy and practice together with a review of the influence that it has had on the construction industry in general.

23.2. The background to Broadgate

The 1980s saw the fundamental questioning of the conventions of office space. Research by Frank Duffy at DEGW (a strategic design consultancy set up by Frank Duffy, Peter Eley, Luigi Giffone and John Worthington in 1973) and others were questioning the standard layouts of office space at the dawn of the Information Technology (IT) era. American practice and requirements were becoming dominant. The Thatcher government introduced the greatest of these changes in 'Big Bang', the deregulation of trading in the City of London in 1986. This in itself brought about a huge building boom in the mid- to late 1980s. Stuart Lipton began his career in property development during the 1960s. He spent the recession years in the early 1980s learning all he could about development and construction techniques used in the US and launched these new ideas in London at the start of the 1980s property boom.

First was the novel Cutlers Gardens scheme completed by Stuart Lipton and Greycoat PLC. This was the forerunner of a series of new ventures that would revolutionise the way London both looked and behaved. New offices with new layouts using new materials and new techniques were being developed above Victoria Station and at 1 Finsbury Avenue, the latter in conjunction with fellow developer Godfrey Bradman of Rosehaugh (to designs by Peter Foggo at Arup Associates). Stuart Lipton liked to establish professional teams, which he used consistently over the years. One of the early innovations was to question the restricted use of structural steel. Encouraged by Lipton, Jim Mathys of the Waterman Partnership followed Stuart Lipton in going over to the US to see these methods at first hand and learn from experience there. The design of Cutlers Court by Renton Howard Wood Levin Partnership (RHWL) was for a five-storey building that reflected the style of Cutlers Gardens opposite with granite elevations and small plain window openings. However, following US practice, Lipton's brief stipulated that no masonry or concrete walls were to be used above first-floor level. The construction solution was to use a composite steel construction: a steel frame of over 300 tonnes with metal deck floors topped by lightweight concrete. To achieve this the London Building Acts had to be rewritten. This was not a simple task as the London Building Acts and the London District Surveyors were particularly conservative. However, Lipton assembled a skilled team to research and argue the case. In the end, Waterman's produced the first draft of the new legislation allowing the introduction of US style steel frames, metal flooring acting as its own fire protection rather than being encased in concrete, metal studs connected the concrete-and-steel slab and the steel beams. The thinner cladding for the exterior – granite slabs precast onto reinforced concrete and later fixed into the structure – was also a precursor of much that would follow in the city.

With this experience Lipton turned to the problem of the contracting industry. Working within the British Property Federation (BPF) and a number of experienced clients they developed the BPF manual (British Property Federation, 1983) which provided detailed and expert criticism of the adversarial and fragmented approach of the UK building industry. It went on to propose a new approach to organizing building work, which took proper account of the client's interests. It also laid the groundwork for many of the construction methods and organizational arrangements used on the Broadgate project.

In 1983, Lipton left Greycoat City Properties PLC, the development company responsible for Number 1 Finsbury Avenue. He formed Stanhope Properties in the same
year and joined with Godfrey Bradman's Rosehaugh to form Rosehaugh Stanhope Developments PLC (RSD), bidding for the Broadgate site. RSD was duly appointed as
development partner to British Rail Property Board in March 1984.

23.3. The Broadgate project

In the formative stages of the Broadgate project the team was advised by Schal Associates, the construction management consultants based in Chicago (Bennett and Gray,
1992). Their track record included the World Trade Centre in New York and the Sears
Tower in Chicago. Lipton encouraged the formation of a company in conjunction with
Tarmac Construction PLC. Schal's role in the early planning of Broadgate was to act
as advisor on the practical details of changes to UK practice that the developers knew
ought to be possible. In many respects American practice was in advance of that in the
UK, particularly in terms of matching project management to the demands made by
the technology being used. RSD used Schal's direct, practical experience of US methods
to demonstrate to the British building industry that the changes that they wanted were
indeed capable of achievement.

The basis of this new approach to construction was established by setting challenging,
world-class objectives for what was, by any standards, a massive and difficult building
project. Broadgate comprised 14 phases, each of which would have constituted a large
building project in its own right. These 14 large building projects were planned to be
completed in just 5 years, prior to Broadgate a typical phase would have taken at least
21 months to construct – RSD set a target of just 12 months for each phase. In addition,
they established stringent cost and quality targets and as if that were not enough, they
were determined that Broadgate should provide a significant contribution to London's
architecture and public amenities.

The required speed of design and construction meant that Broadgate provided a
major challenge to the capacity of the British building industry. Many innovations had
to be used, but to avoid new ideas leading to random and disruptive change, they had
to be introduced in a controlled and systematic manner. It was clear from the outset
that completing this major task required a quality of construction management not
commonly found on British building projects. However, Broadgate had two crucial
factors working in its favour. Firstly, the 14 distinct phases provided an important opportunity to create an organization with a reasonably long life, which had time to learn
from repeating essentially similar activities. Secondly, Broadgate had, in RSD, a client
who provided clear leadership for this great enterprise. To provide the management
capacity, RSD brought together Bovis PLC and the formative Schal in a joint venture
after phase 4. In practice, Stanhope took the lead in the practical execution of the
project, which allowed the development of the technical and management processes
in a consistent way.

The developer's brief was the first step in the RSD approach to provide their design
and management consultants with a clear brief that describes the requirements of the
particular phase; the standards which must be applied in designing, manufacturing
and constructing the building; and the services to be provided by each consultant.
All of these instructions to the consultants were given in unusual detail. The brief

described the overall purposes of the building, its key dimensions and layouts, and the type of facilities to be provided. It provided detailed specifications and practical advice on designing each and every element of the finished building. All of this was based on the use of technologies, which RSD knew would provide the quality, speed and economy that they demanded. Importantly many requirements of the brief were based on ideas devised with the help of the design consultants and specialist contractors systematically recorded from previous projects. However, the brief also required that teams continued to review and improve on this reservoir of good experience. This use of previous experience is often talked about but this was one of the first examples of it being implemented to good effect.

The second step in the RSD approach was to select and engage talented consultants to design and manage their projects. They looked for multidiscipline design consultants, or those experienced in working together in multidisciplinary teams, capable of working with the construction manager. Multidiscipline work in well-established teams enables rapid, almost telepathic, communication between the separate disciplines needed in the design of complex, modern buildings. It is much faster and more efficient than working through regular formal meetings and communicating by telephone, post or fax.

On Broadgate, RSD used the construction management approach because of the greater flexibility that it provided to bring talented people from many companies into an integrated team. This was an American approach, which recognized the need for construction knowledge and experience to be involved from the earliest stages of major building projects. The construction manager was employed directly by the client as a consultant. Having decided on the overall management framework, RSD then searched for the best available people to undertake all the demanding tasks required by a project of the scale of Broadgate. Talented people were recruited from all over the UK and beyond. The main requirements were deep experience in the technical competence required to undertake a particular task and enthusiasm for the idea of joining a team committed to searching relentlessly for better, faster, safer and more efficient methods.

Furthermore, the construction management approach allowed involvement by RSD and enabled them to act as a catalyst within the team. This is where the right attitude is engendered and relationships with contractors and the Unions fostered: the contractors obtained 70–80% repeat work through Broadgate and future developments.

The Broadgate team for phases 1–4 was led by three remarkable men. Firstly, Peter Rogers, the Construction Director, who provided the overall driving force for the whole project team. He subjected himself to a working day, which typically chaired a dozen or more meetings stretching over 14 or more working hours. His style was to constantly try for 10% extra quality, speed and productivity. He embodied the 'can-do' attitude of solving problems as and when they arose which came to characterise the Broadgate project. Secondly, Peter Foggo of Arup Associates led the design team. The physical results of Broadgate phase 1–4 provide what many informed commentators regard as one of the triumphs of modern architecture in London. The buildings and the public places they provide are a tour de force in architecture based on a machine made, high-technology product. To achieve the depth of quality, which characterise his work, Foggo insisted on being involved in the design of every detail. The third man in Broadgate's top team proved very able to meet the challenges provided by Rogers and Foggo. He was Ian Macpherson, who headed the Bovis construction management team. He was already deeply experienced in Bovis' well-established management methods when he was appointed as project director for Broadgate. The Bovis team provided a range of management services including project management, value engineering, cost planning and control, programme control, dealing with tenders from specialist contractors and managing the construction. It was Macpherson's eagerness to question established methods and his willingness to listen to ideas from anybody and everybody in the team, which made him the perfect foil for Rogers and Foggo. The intense questioning of every aspect of normal practice helped turn the first four phases of Broadgate into what was, in effect, a construction management laboratory. Management methods were developed, refined and tested in the search for a way of combining Foggo's design flair with the high management efficiency demanded by Rogers.

By phase 11 of Broadgate, the management methods had settled into a smooth, efficient pattern, which set the standards for construction management in the UK for the foreseeable future. Indeed, the methods developed on Broadgate have been influential in shaping the views of the Construction Management Forum whose report published by the University of Reading's Centre for Strategic Studies in Construction (Cornick *et al.*, 1991) provides an authoritative statement of current good construction management practice for the UK.

The third main step in the approach used at Broadgate was to involve specialist contractors sufficiently early in the design of each building to enable their expertise to be devoted fully to meeting RSD's objectives. Such early involvement recognized that modern building technology is provided, not by general contractors, but by specialist contractors who design, manufacture in factories and then construct on site very sophisticated elements of buildings. RSD welcomed and encouraged specialist contractors, securing the maximum input from them and thereby including them in the team; the incentive being repeat business.

Broadgate was a unique development because of the seriousness with which two major strands of research were used. Firstly, research into what the users of the buildings want and secondly, how buildings can be constructed better and more efficiently to meet these requirements. Broadgate would not have been constructed so quickly and directly without major investment in research into construction methods. RSD's approach to construction management led to major technological improvements that made the buildings not only extremely quick to build but also cost effective and of a quality that stands comparison with office developments anywhere else in the world. Peter Rogers and members of the Stanhope construction team, in order to push technological development, joined a large number of technical committees across many specialist sectors. This heavy time commitment helped keep the development of new products and technologies very focused to ensure they met Stanhope's need to keep driving radical change.

23.4. Influence following Broadgate

At the height of the 1980s development boom Rosehaugh and Stanhope stood out as beacons of dynamic change. Broadgate as a development was highly praised and it was from this base that Lipton and Rogers developed a public career to foster and promote what they knew could be achieved. Lipton focused on architecture and design and Rogers on construction performance. Sir Stuart Lipton was appointed as CABE's first Chairman in May 1999. When the government decided to set up the Commission for Architecture and the Built Environment to promote good architecture, planning and regeneration; he was the natural choice as Chairman. From scratch, under his stewardship, CABE established itself as a widely respected and hugely influential body always pushing the best and fearlessly criticising the mediocre. Tessa Jowell said on his reappointment:

'The difference that good architectural design can make to improve the lives of ordinary people and to deliver 'liveability' is at the heart of CABE's work. Sir Stuart Lipton's knowledge, enthusiasm and experience have made him a strong and imaginative leader. He is passionate about the importance of good design and he enjoys great respect in the development community and built environment professions. He has served CABE with distinction and will play a crucial role in CABE's continuing development. He has established CABE on foundations that will stand the test of time, and his experience and expertise will enable CABE to meet the exciting challenges it faces in the future'.

An Honorary Fellow of the Royal Institute of British Architects, Sir Stuart's knowledge of architecture and the built environment is demonstrated by the positions he has held in several architectural bodies. As well as serving as a member of the former Royal Fine Art Commission from 1988 to 1999, Sir Stuart has also been a Trustee of the Architecture Foundation since 1991 and served as a board member of the Royal National Theatre from 1988 to 1998. He is a board member of the Royal Opera House with responsibility for its development. Sir Stuart has been involved in a number of arts buildings including the Glyndebourne Opera House, the Sackler Wing of the Royal Academy, Tate Modern and the Sainsbury Wing of the National Gallery.

Sir Stuart was also perhaps the first developer to be truly interested in the construction of buildings and to understand the importance of the construction process to good development. He can debate knowledgeably double handling of materials and hook time with the best logistics contractor. Sir Stuart has shaped large parts of the environment in which we work – Ludgate, MidCity Place, Paternoster, Chiswick Park and the Treasury Building have all been developed under his or Stanhope's guidance (British Council of Offices, 2006).

Peter Rogers has remained at the operational level albeit from a client's perspective. The Stanhope business evolved in part as an expert client, which is really the stage before project management. However, because of their expertise Stanhope can take an integrated view of the development problem and debate any and every aspect. Rogers has therefore provided, with his colleagues a service to clients on many of the arts buildings listed above as well as continuing to promote change from the Stanhope perspective. He encouraged the setting up of Asite.com for online procurement and the East London Consolidation Centre for logistics management to London developments. For a while he was chairman of the Strategic Forum for Construction and then Constructing Excellence. Both of these organisations have been the drivers for the implementation of the Latham and Egan reports. His latest project is as chairman of the Construction Industry Council's 2012 Olympics task group. Here he has set an agenda for change to the whole industry by 2012 called the 2012 Olympic Construction Commitments. It is an agenda for simplified good practice for clients, contractors and suppliers where the Olympics can be used as an exemplar for change and to illustrate the effectiveness of good practice.

23.5. Conclusions

The 1980s were a period of radical change in construction. This was consolidated in the 1990s. Clients in the UK such as Sainsbury, Marks and Spencer and Tesco and developers such as Slough Estates led the way in consolidating the lessons learnt. However, one developer, Stanhope, has stood out from the field. Sir John Egan interviewed Sir Stuart Lipton when he took over at the British Airways Authority (BAA) and the lessons learnt at BAA found their way into the Egan report. Lipton and Rogers have been extremely influential by challenging the industry, doing better themselves and then asking for more. To remain competitive, they have adapted and adopted research, new ideas and new practices. Originally the new practices were from overseas, but more recently home-grown developments are leading the way. Lipton and Rogers have had to lead the changes in the industry in order to get the teams that they want to get the performance that they wanted. Others have observed and the last 10 years has seen a massive growth in capability as people, influenced by these two, have had the confidence to try themselves. Stanhope has continued to work closely with Bovis (now Bovis Lend Lease) as a method of consolidating the expertise. Macpherson left with a large number of the Broadgate team to form MACE, which has lead the further development of construction management. In fact, many of the most influential people and clients currently in the UK can track back their experience to this amazing development or been influenced by Lipton and Rogers. Rarely, however, has the same performance been achieved, so consistently. Stanhope took possession of their projects in a way through equity stakes that raised them above consultants to stakeholders and this gave them the drive and authority to push the change and still push for change.

References

Bennett, J. and Gray, C. (1992) The construction of Broadgate. In: *The History of Broadgate* (ed P. Hunting). London, Stanhope, pp. 48–60.

British Council of Offices (2006) Presidents valedictory speech on the award of the Presidents Medal to Sir Stuart Lipton.

British Property Federation (BPF) (1983) *Manual of the BPF System for Building Design and Construction.* London, British Property Federation.

Cornick, T., Bennett, J. and Murray, J. (1991) *Construction Management Forum – Report and Guidance (1991).* Centre for Strategic Studies in Construction, University of Reading, Whiteknights.

Flanagan, R., Goodacre, P., Gray, C., McLaughlin, N. and Norman. G. (1979) *UK and US Industries – A Comparison of Design and Contract Procedures.* London, Royal Institution of Chartered Surveyors (RICS).

24 Customer focus: time, the enemy of desire – a contractor developer perspective

Chris Woods

24.1. Introduction

Historically, the reputation of the construction industry and its 'builders' was the source of jokes, ridicule and despair to the prospective clients of the industry. Ten years ago, Wates started on a process to shift from an adversarial contracting proposition and to focus on improving delivery of its products and services to its customers. Despite early success, it was only in 2001 that Wates started to really understand that they needed to see what they were delivering from the perspective of the customer rather than from an internal viewpoint. Customer focus is a never-ending journey and now drives all Wates does. However, there appears to be one issue that stands between Wates and total customer satisfaction, namely the time taken from inception to delivery to produce the product.

24.2. What enables customer satisfaction?

There is general agreement by most customers that there are three things that are fundamental for customer satisfaction – to time, to budget and to quality or specification. It is clear though that there is more than this. The majority of customers select their construction suppliers on quite a number of other criteria. Cultural fit, attitude to safety, attitude to sustainability, attitude to off-site and ability to deliver innovation; all of these can be described as aspects of the ability to deliver added value and specifically more of what is most valuable to the customer's business. More content that is valuable for the same or less money.

24.3. What prevents customer satisfaction?

If the customer does get exactly what they have asked for, to time, to budget and to quality, it is clear that the customer will not be satisfied particularly if increases in cost and programme are not the accountability of the construction team but due to inadequacies in the original brief that necessitates changes to the design – either during the construction period or during the period immediately post-procurement.

Delivery period days

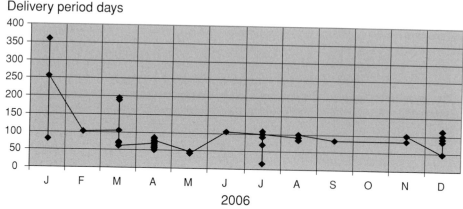

Figure 24.1   Delivery periods – M&S stores.

A critical issue that also hinders the process is an aspect that is becoming increasingly evident, the customer and sometimes even the customer representative cannot precisely read drawings. When they are signed off for construction, the customer is unaware exactly what the finished article will be like in reality. The only people Wates research shows who can visualise a drawing as a three-dimensional reality are people who have been trained to draw and read the highly stylised drawing system of representing buildings. For someone who has been trained in this language it is unimaginable that others are not adept at reading them. This failure usually comes to light when the building starts to take shape and the customer realises changes are needed, by then it is too late to achieve the highest levels of customer satisfaction.

However, even if all the aspects of delivery to time, cost, quality and added value are achieved and the building owner and other significant stakeholders can read and sign off drawings, i.e. delivered exactly to brief, it may still not be what is needed. This is due to the fact that the design will have been done months, if not years before the delivery date and even the most clear-sighted customers are unable to predict exactly what they will want at the delivery date, hence imperfect customer satisfaction. Fundamentally this is a failure of process. All customers must attempt to make an educated guess as to what they will need a number of years or months down the line. It looks as though it is never possible to deliver very high levels of customer satisfaction if the delivery of the product is anything other than a few weeks or months after it has been ordered. However, the subsequent brief history of changes to procurement practices of buildings might illuminate routes that mitigate this flaw in delivery.

In 2006, Wates Group delivered 51 Marks and Spencer (M&S) 'Simply Food' stores within the UK. Over that year the Wates Retail division improved delivery and in time all stores were delivered in less than 3 months (see Figure 24.1). Wates also worked on improving processes to reduce the lead in time with M&S. Just after the completion of the stores M&S announced their decision to go zero carbon by 2012. From the perspective of hindsight this might look like stores were out of date within a year of delivery reducing customer satisfaction however the right decisions were made at the right time. What is true is that the shorter the delivery period decreasing the time between briefing decisions and delivery of product would increase the likelihood of product being what the customer needs for its future assets.

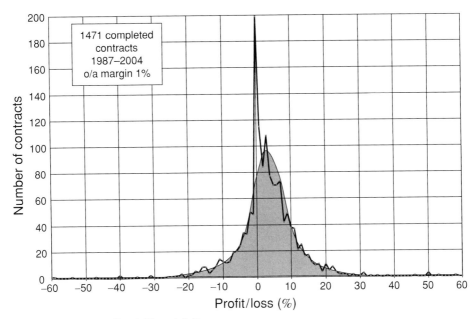

Figure 24.2 Unpredictability of delivery.

24.4. Unpredictability of delivery

There are processes that attempt to mitigate the inevitable customer disappointment of not receiving exactly what is wanted at the time of delivery – the variation process. This process does have its limitations though because it is highly likely to change both the cost and delivery date adversely, potentially violating at least two of the basic customer requirements: budget and programme. One aspect that is not understood by many people is just how unpredictable delivery of construction really is. Construction is not like post-prototype manufacturing where it can be precisely calculated exactly what something will cost. There is no prototype and the variables that can influence costs are too great in number to have predictable certainty. Across our industry about 30% of all projects lose money (see Figure 24.2).

These performance graphs are similar for the customer. However, there is no reverse correlation between profit and loss for customer and contractor. It tends to be if the customer obtains a building within budget the constructor makes a profit; if the customer receives a building outside of budget there is a greater likelihood that all parties will make a loss. There appear to be specific tendencies that will push projects either to the right or left of the bell curve (see Figure 24.2), for example, project team familiarity, contract type, repeat customer, supply chain constancy, rate of construction and speed.

24.5. Programme implications

Certain contract periods produce greater uncertainty in results. Project speed as indicated above can influence customer satisfaction. However, it also influences profitability. Very short contract periods up to 5 months tend to increase the overall likelihood

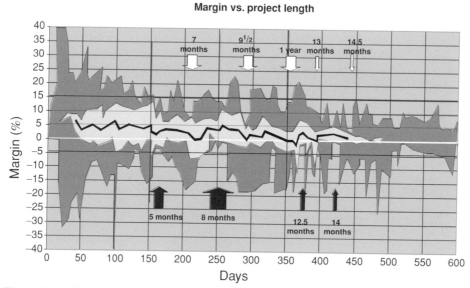

Figure 24.3   Programme uncertainty and profit.

of making money, however, it also brings with it higher uncertainty and increased risk (see Figure 24.3).

24.6. Process improvement over time

Various changes throughout history have occurred to the process of procuring the delivery of buildings. All improvements are about attempts to give the customer more of what they want and generally they attempt to reduce time between final design decisions and build delivery to improve customer satisfaction. Different procurement methods and how they have impacted on innovation and customer satisfaction are discussed below:

- Vernacular
- Pre-designed
- Traditional design, flexibility and variations
- Design and build
- Two stage
- Frameworks

These procurement routes could be regarded as mitigation strategies to deal with risk and particularly the potential failure to deliver customer value. Different approaches result in different levels and approaches to innovation. Failures in organisational intelligence can cause problems for innovation even without the complexity of having fragmented project teams. Understanding customer value can be hard for even a customer with potentially different departments, asset management, building procurement and sales. It is even harder for building procurers or project managers, designers and constructors who are a further step removed from customer needs.

24.6.1. Customer as deliverer

When the customer was also the builder then customer satisfaction and innovation tends to be very high. For example, look at Skara Brae, a Neolithic village on the islands of Orkney, UK, constructed 4000 years ago. As the customer was the deliverer there was direct innovation pull, producing exactly what the customer needed with the resources available. The drivers were clear though, very little wood, a cold climate and large quantities of shell midden, the waste product of their diet. They produced insulated double skin external walls with a U value of 0.45 w m^{-2} – something the UK construction industry did not create until building regulations forced them to in the 1980s. Due to intimate knowledge of the properties of the natural laminar slate the customer/deliverer also created the damp proof course, the street/corridor, sliding doors and what was clearly a sustainable community thousands of years before it was rediscovered in the UK. The act of the building process was also part of what communities did. This in itself creates an aspect of customer satisfaction.

24.6.2. Vernacular

Construction 600+ years ago was relatively simple – 'I will have one like that one', or one like that one but with definitive exceptions, was how over 99% of buildings were procured. These buildings were produced relatively quickly and cheaply and they gained high levels of customer satisfaction. Buildings would be amended as and when with virtually no administrative hurdles. The innovation was an aspect of slow evolution over many projects. The process gave rise to specific local fashions mainly due to local material usage that some designers still try to replicate today. Using existing buildings as prototype models for new ones resulted in virtually zero design time; construction, therefore, could start very quickly. As there was usually a single customer who was part of the process there was high customer satisfaction. However, with a few buildings such as churches or castles construction times were occasionally generational. Producing different issues as the culture of the initiator was not necessarily the same as the generation that received the finished construction.

In the 18th century when the building code arrived, engineers and later architects attempted to mitigate failure in customer satisfaction by producing various designs prior to construction. Innovation started to move away from the construction site to the drawing board.

24.6.3. Pre-designed

The building code helped created a kit of parts thinking for Georgian (UK) buildings with innovations like 'coade' ornamentation for different appearances of buildings. The design phase was quick, with simple building types, and simple drawings, i.e. section, plan and elevation. Customers could get what they were prepared to pay for and were generally happy, hence the increasing number of designers. Fashions changed relatively slowly and changes could be relatively easily incorporated into designs. There was always a clear choice. If an existing building did not suit needs it was easy to see where it was desirable to build new. Everything was generally cost

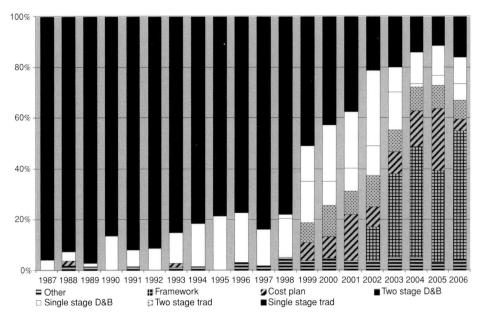

Figure 24.4   Changes to procurement route.

plus and relatively transparent. In the majority of situations innovation occurred in the period of translation between 'I will have one of those except I want it to do this or be for this function'. Representing buildings on paper, pre-construction, allowed customers to see to a degree what they were going to receive as a building. However, innovation on paper produced massive uncertainty in delivery. Innovation could easily result in major cost increases and delays in delivery both resulting in reduced customer satisfaction. This was a golden age for innovations. Many innovations were the result of the rapid developments in new materials, in the use of iron, steel, glass and concrete. Little was needed to be done to innovate the organisational process of design and construction. The designer/constructor built the building and so the later disconnects between costing, design and construction never hindered the process of innovation to deliver what the customer wanted. Cost and programme overruns and failures in quality caused major changes to procurement processes in the 20th century.

24.6.4. Traditional design, flexibility and variations

Even 10 years ago the vast majority of projects were still traditional where the intention was to complete design drawings prior to start on-site (see Figure 24.4). Design development and any innovation occurred within the pre-contact stage; the downside of this was that the length of the design process was beyond the period of time that the customer could envision the required delivered product. From the 1960s onwards this tended to cause a new element in the brief; customers wanted flexibility. The concept of flexibility is complex, meaning different things to different stakeholders. To an architect it might mean the ability to produce a modern rather than a traditional load-bearing masonry structure, to others it might mean extendibility. However, to customer, it usually meant that the customer was not exactly sure what was wanted at the point of delivery. Customers ended up paying a premium for so-called flexibility when all it really meant was that the design would change during both the design and construction periods. Rarely are buildings that had 'flexible' designs ever changed post-construction. It is clear from the extent of refurbishment work that is done by Wates that little except the fundamental structure is generally preserved. When pre-1960s buildings are changed on many projects it is just the facade that is retained. Ironically, buildings that were designed to be changed are generally demolished usually because extra floors are needed that the structure was not designed to cope with. A customer's real flexibility usually means change to designs during construction. This creates its own problems in other areas of customer satisfaction, namely cost and programme. Innovation also causes problems for traditional contracts, as failure to fully understand the required product at the start tends to produce additional changes within the design during the build. Due to disconnect between the design and construction process within traditional contracts, innovations generally create higher levels of uncertainty and it is found that they also create a mixed reaction in customer satisfaction, particularly if changes result in cost or programme overruns.

24.6.5. Design and build

Over the last 10 years there has been an increase in the prevalence of new procurement forms. These have been introduced as ways of transferring fiscal risk. However, a by-product has generally been higher levels of customer satisfaction due to the reduction in time between briefing and delivery. Analysis of Wates projects shows that this process effectively allows a second design-briefing period at tender stage, which is inevitably closer to the delivery date. It results in customer desires being more accurately represented in the final product. Greater interaction between constructor and customer improves the sharing of knowledge at a practical level allowing inclusion of desires with a higher possibility of controlling cost and programme. Innovation is limited, however, due to the customer requirement to have competitive tender process for the appointment of sub-contractors; this limits the early involvement of sub-contractors during the design phase. During the last 10 years Wates have found that traditional contracts have also changed to include more contractor design portion elements. This tends to improve the delivery of higher customer value and is also a way of reducing the time between briefing and delivery; many traditional contracts have at least 50% of the contract value as contractor design portions.

24.6.6. Two stage

Since 1998, analysis of Wates procurement routes have seen a steady increase in the development of two stage contracts, first within traditional contracts and then in design and build (see Figure 24.4). The development of the design is done together by the whole design and delivery teams and generally at a later stage. Cost programme and functionality can be considered together to produce higher certainty in these three areas. A greater level of customer knowledge is shared throughout the whole team. Innovation and flexibility is usually more appropriate to customer need and there is less disruption to the construction process due to the greater control over the total delivery.

Wates Group turnover is generally 1% of total UK construction turnover excluding heavy civil engineering projects that the company do not carry out. Wates do not have sufficient data on all contract forms to know if this is representative of all UK construction although experience from Wates staff that have worked for different main contractors suggests that the changes in the prevalence of different procurement methods is similar for the whole of the UK. Geographically, the incidence of newer procurement forms tends to occur in the Southeast of England first and has a 2–3 year lag as you go further north of the UK. Having said that, geographic time lags have not occurred with the introduction of frameworks and partnering contacts.

24.6.7. Frameworks

Framework contracting developed out of retail procurement methodology in the late 1990s and eventually spread into the public sector by 2001. Main contractors and principle designers are pre-selected for periods of 5–7 years to carry out a percentage of work for serial building procurers usually on the basis of overhead and profit levels together with value-adding ability. The development of framework contracting has had numerous benefits in the delivery of higher levels of customer satisfaction. The improved customer knowledge has resulted from getting to know more aspects of customer organisations not just the estates and asset managers but also users and the customer's customer. Supply chains have become more integrated and are starting to be extended beyond tier one and two organisations, of the contractor and designer to now include principle sub-contractors.

The possibility of having real customer pull in research and development (R&D) has resulted in changes in R&D away from supplier push and component development to the full inclusion of the supply chain in assembly level product and process innovation over the life of frameworks.

Other innovations include the possibility of designing projects to a cost rather than the historical process of costing designs. It has also resulted in improved predictability for both budget and time. Shortened periods between design and delivery have their own benefit of being able to deliver more of what the customer needs at the time of delivery. It returns the process back to the customer as they are part of the process. There is one further aspect that customer focus requires and that is repeat and or serial customers getting to understand what and how to deliver to particular customers is critical to profitability. Figures 24.5 and 24.6 illustrate the benefit to Wates of repeat customer as opposed to one-off customers.

24.7. Customer focus

Customer focus is a journey and understanding that the majority of construction customers have other stakeholders who are also necessary to satisfy is key. However, predictable delivery on its own will not be sufficient to satisfy customer desires. Managing stakeholder expectation must change to delivering stakeholder expectation. Understanding those expectations will also never be enough due to the realisation that

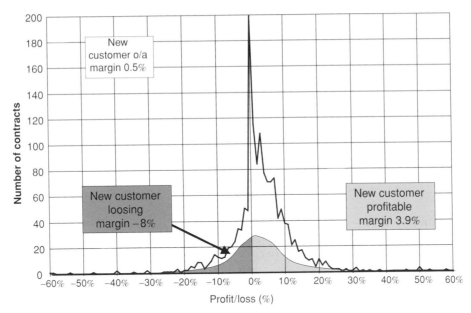

Figure 24.5 Margin – new customers.

expectations naturally change through time just due to time itself. For years constructors and designers have mistakenly thought that customers value the process of construction, its complexity, the dirt, the danger and the disruption – they do not. If construction is to get to really high levels of satisfaction, the industry must move towards concepts of invisible construction, towards zero time and whole buildings as product.

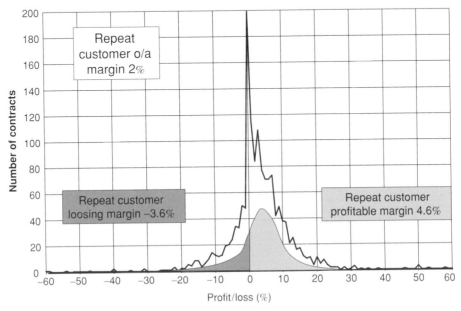

Figure 24.6 Margin – repeat customers.

Even then the industry will only be half way there; buildings are not just products – they are vehicles to allow customers and wider stakeholder communities to increase the value of living, working or playing in those spaces. Understanding customer value is complex. You only definitely know what does or does not work after it has been delivered – when it is too late. Post-occupancy assessment and evaluation is critical for the development of buildings as products. Construction has not yet achieved its Model T Ford stage in delivering customer value. The industry still has the equivalent of the red flag in front of our vehicle in the town planning process that prevents near instant delivery to customer need.

A way to potentially rethink some of the issues concerning time as a fundamental obstacle to the delivery of what the customer actually wants at the time when a facility is delivered is to rethink the way the industry conceives of buildings. There are three essentially different elements to buildings, each with different life spans:

1. The structure with a life span of at least 100–200 years.
2. The external fabric with a useful life span of 50–100 years.
3. The internal fit-out with a fashionable life span of 10–20 years.

In non-traditional construction these three elements are increasingly becoming less integrated. Not only do the three elements of construction have different life spans, but they also have different design and construction periods. In order to truly satisfy the customer delivery must be within a time period that is within the ability of the customer to know exactly what is required at the point of delivery. To avoid extras to a contract and the potential of increases in the delivery period variations need to be eliminated. This is at odds with the customer needs of functional delivery. In Wates' experience customers cannot visualise the final building outcome and this gives rise to many variations. Further variations occur due to the functional brief changing due to the inability to foresee requirements at the point of delivery. The inability to visualise can be overcome by the use of three- and four-dimensional visualisation tools, but the inability to foresee needs can only be overcome by decreasing the time between briefing and delivery. Some customers wishing to avoid the disruption to the construction process and unbudgeted time and cost variances to happen allow contract completion to occur and then they embark upon post-delivery changes. In Wates' experience 6 months is the maximum period that customers can foresee exactly their requirements. Due to the inherent uncertainty of the planning process the delivery of both the design and construction of a whole building within a 6 month period is extremely unlikely giving rise to less than perfect customer satisfaction. To this end Wates are finding that some customers mostly in the retail and commercial sectors split projects into separate contracts, shell and core, and then fit out as a totally separate contract sometimes using specialists. The fit out elements are not hampered by the planning process and, therefore, this helps the delivery period to be within the 6 month period of possible foresight and therefore improves customer satisfaction. However, it is less than satisfactory should the fit out requirements require changes to the shell and core elements.

Off-site delivery processes can aid simultaneous construction and reduce time periods. However, what is really critical for successful delivery with multiple building procurers is the development of standard building types that can be particularised at short notice. Increasing the separation of the different hierarchies of elements is also helping to improve customer satisfaction. These kinds of processes are what Wates Group has been developing for a number of framework customers and hark back to the period of 'we will have one of those'. Being able to gain generic planning approval is still the principle obstacle that is preventing total customer satisfaction.

24.8. Conclusion

Construction must deliver higher levels of customer satisfaction, this will only follow from ever-increasing levels of customer focus by integrated and fully participating construction and design teams. This will drive innovation and will help the industry overcome the ultimate barrier to reaching the highest levels of customer desire – the time taken to design and deliver a project.

25   The role of the client in building site innovations

Frédéric Bougrain

25.1. Introduction

In economic literature pertaining to the construction industry, the client often appears as dissatisfied. Inability to deliver a project on time, to keep to the quoted price and poor performance of the final building, are the main criticisms advanced by clients (Egan, 1998). Paradoxically the client also appears as one of the main sources for changes. Several reports published in the UK and in Australia consider that clients 'who know what they want and how much they are prepared to pay for it . . . ' can spur innovation (Egan, 1998, p. 33). This move appears in the visions of the managers of Australia's property and construction industry. They 'challenged clients to take a more proactive and educated leadership role and to enforce a total life cycle cost approach' (Hampson and Brandon, 2004, p. 14).

In the French building and construction industry, several criticisms emanate from the client (called the 'Master of the work') while, at the same time, the contractors criticise the clients for selecting them on the basis of lowest cost. While the private client has considerable freedom to select his project team, the public client has to behave according the law no. 85.704 laid down on 12th of July 1985. The public client, who is not presumed to know anything about construction, has to be surrounded by professional advisers in order to comply with the requirement for public accountability. Under this law, the public client generally establishes two different contracts with the architect and the contractor. The new law towards public private partnership, enacted in June 2004, proposes a complementary framework. Under this new scheme, design, build, finance and operation are transferred to private sector partners. This change is expected to encourage more innovation because of the competitive dialogue and because the public sector remains the client throughout the life of the contract.

This chapter aims to examine how clients are involved during the innovation process. To achieve this goal and to avoid any confusion, the definition of the client as adopted by CIB Task Group 58 will be used: 'A client is a person or organization, who at a particular point in time, has the power to initiate and implement design and construction activity to improve the performance of an organisation's social or business objectives'.

The first and second sections define innovation and looks at the role of clients in the innovation process from a literature survey. The last section draws upon case studies of 64 contractors recently recognised for developing innovative products and methods by a jury of people working for the building industry. The nature and origin of the innovations identified and the involvement of the clients during the innovation process are analysed.

25.2. Innovation: some definitions

Most of the studies undertaken to examine innovation focuses on technological innovation and particularly product innovation and process innovation. Service, marketing and organisational innovations are often neglected. But several books (e.g. Sundbo, 1998; Metcalfe and Miles, 2000; Gallouj, 2002) and papers (Sirilli and Evangelista, 1998; Drejer, 2004) show that innovation is broader and that something is missing in the current debate. The third edition of the Oslo Manual (OECD, 2005, p. 46) identifies this lack and addresses the question of non-technological innovation:

> An innovation is the implementation of a new or significantly improved (good or service), or process, a new marketing method, or a new organisational method in business practices, workplace organisation or external relations. (...) The minimum requirement for an innovation is that the product, process, marketing method or organisational method must be new (or significantly improved) to the firm. This includes products, processes and methods that firms are the first to develop and those that have been adopted from other firms or organisations.

As indicated by the Organisation for Economic Cooperation and Development (OECD), overlapping between products, process, marketing and organisational innovations may exist. It is mostly at the interface between process and organisational innovations that mentions of this sort appear. Indeed, 'both types of innovation attempt – among other things – to decrease costs through new and more efficient concepts of production, delivery and internal organisation' (OECD, 2005, p. 55). Following the OECD guidelines, we will consider that 'process innovations deal mainly with the implementation of new equipment, software and specific techniques or procedures, while organisational innovations deal primarily with people and the organisation of work' (OECD, 2005, p. 55).

25.3. Is the client driving the innovation process of construction firms?

In the economic literature many studies have been devoted to the determinants of innovation. The approaches defending the 'technology push' hypotheses (in this case the impulse comes from the supply side) have been traditionally opposed to those focusing on the 'demand pull' (the origin comes from the demand side). According to Mowery and Rosenberg (1979, p. 153), 'any careful study of the history of an innovation is likely to reveal a characteristically iterative process in which both demand and supply forces are responded to'. Lundvall (1988, p. 358) also notes that 'many innovations, appearing as purely supply determined, have their roots in a user–producer interaction placed early in the chain of innovation'.

This idea of interaction lies behind the notion of clients driving innovation. The client does not only instigate the innovation process, but also pilots the process and involves the other stakeholders in the making of the solution. Consequently, there is a co-production between the client and the other stakeholders.

A short literature survey indicates that in the construction industry the client does not always drive the innovation process. Pavitt (1984), who was the first to emphasise the existence of sectoral patterns of technical change, did not deny the role of the client but he emphasised the role of supplier. He categorised general contractors as 'supplier dominated firms'. Firms from this category devote few resources to Research and Development (R&D). They focus their innovative activities on processes. 'Most innovations come from suppliers of equipment and materials, although in some cases large customers and government-financed research and extension services also make a contribution' (Pavitt, 1984, p. 356).

Many innovations introduced by contractors are not influenced by the client. General contractors tend to focus their resources on the effective management of the building site, which can be considered as their core activity. Most innovations aim at circumventing bottlenecks, which enables improvements in the productivity and safety of the building site. These innovations, mainly informal, come from people working on the building site. These people have to solve recurring problems that disrupt their daily activity (Bougrain, 2003).

Several reports (Egan, 1998; Hampson and Brandon, 2004; Barrett and Lee, 2005) considered that the construction industry needs sophisticated clients and urged the public sector to play a leading role. Manley *et al.* (2005) put forward some arguments to explain why public clients are very often risk-averse. They suggest that public clients are not ready to support innovative solutions that are often more costly and provide unknown results. Such conservative behaviour was also identified by the National Audit Office (2001). They suggested that some private finance initiative (PFI) projects do not always lead to innovative solutions because public clients are not always receptive and have fixed views on design features and matters concerning service delivery.

However, it appears that the client also plays a positive role. The Building Research Innovation Technology and Environment (BRITE) survey (Manley, 2006) indicates that leading and experienced clients often provide opportunities for innovation. This is particularly the case for public clients who tend to favour value-based tender selection:

> Sixty percent of the 383 survey respondents nominated repeat public-sector clients as 'encouragers' of innovation in the industry. The survey also found that, compared to other industry groups, such clients had the highest rate of investment in research and development, the highest rate of adoption of advanced practices and technologies, the best return on innovation and they were ranked fifth among 14 sources of ideas listed in the survey.
>
> Manley, 2006, p. 5

Gann and Salter (2000) also considered that large clients can drive the innovation process. This is the case for owner-operators of large facilities who can put high pressure on the construction team and force them to deliver the building on time and achieve good quality construction. 'They also wish to improve lifecycle performance characteristics and enhance flexibility to meet unforeseen changes in demand' (Gann and Salter, 2000, p. 961).

Brandon (2005) estimates that clients need intermediate agents (such as designers) who define new solutions that express the client's needs. But these studies presented by Manley (2006), Brandon (2005) and Gann and Salter (2000) are focused on large and exceptional projects. Innovation is stimulated because clients preferred value-based tender selection and relied on strong project teams.

According to Sexton and Barrett (2003) who focused on small construction firms (which are dominant in the industry), the role of the client varies according to the market positioning of the contractor. Most small construction firms react to external changes in their business environment and innovate under the pressure of their environment (such as changing client needs and unpredictable project-specific conditions). But the modes of innovation may differ from one company to the other. Under Mode 1, the firm has a limited influence on its environment. Innovation is driven by 'cost-oriented relationships between the client and the firm' (Sexton and Barrett, 2003, p. 629). Under Mode 2, the firm is proactive. It has the ability to influence its external environment because of its innovation capabilities (its organisation and its human resources). 'Mode 2 innovation concentrates on progressing multiple project, value-oriented relationships between client and the firm' (Sexton and Barrett, 2003, p. 629). It seems that Mode 2 enables the development of fruitful interactions between contractors and their clients.

It appears from this brief literature study that the role of clients during the innovation process is rather ambiguous. The nature of the innovation and the nature of the project together with the internal capability of the contractor (ability to interact and to interpret the client needs and benefit from the client demands) are key elements that influence relationships between contractors and clients.

The next section examines the role of the client in building site innovations. The role of the clients is evaluated by drawing upon case studies of contractors recently awarded for developing innovative products and methods.

25.4. Methodology of the empirical study

In 2000, a National Innovation Award concerning construction firms was launched in France. Every 2 years, since this date, companies are invited to present innovations that have been successfully implemented. Applications are judged by a jury of people working for the Ministry of Housing, the national federation of contractors, the federation of small contractors, the French innovation agency, Centre Scientifique et Technique du Bâtiment (CSTB), a journal dedicated to the building and construction industry and a regional delegation from the Ministry of Construction.

The aim of the committee that organises the awards is not to be elitist. One of its purposes is to transform innovation from tacit into codified knowledge[1]. It also aspires to promote the image of the construction industry and the diffusion of innovative approaches. Indeed construction is often characterised by its inability to learn from one project to the next. Winch (1998, p. 271) mentioned that in the UK there is an 'exploration trap where technologies are continually re-invented in a circular rather than progressive manner (...)'. Gann and Salter (2000) considered that construction firms tend to re-invent the wheel[2]. Slaughter (1993) also indicated the problem that occurs among builders who use stressed-skin panels which results in duplication of effort.

Each firm to which an award has been made has its name published in the French weekly construction magazine 'Le Moniteur'. This journal also describes in a few lines the main characteristics of the 'best' innovations and by these means firms can learn from one another.

In 2003, CSTB was commissioned to examine the impact of the awards granted in 2000 and 2002. Firms were questioned about the origin and the diffusion of the innovation, the organisation of the innovation process, the results of the innovation and the impact of the award. Sixty-three face-to-face and telephone interviews with general managers of small companies and technical executives of large groups were carried out by CSTB. This represented 77 innovations awarded out of 97. The difference was mainly caused

Table 25.1 Categories of the innovations awarded

	Categories of innovations				
	Building techniques	Safety and work conditions	Methods and organisational schemes	Environmental approaches	Total
Applications received in 2000	39	28	13	6	86
Applications awarded in 2000	8	8	4	6	26
Applications received in 2002	63	87	38	10	198
Applications awarded in 2002	14	25	17	15	71

by the unavailability of executives and the bankruptcy of some construction companies. In four cases, enough information about the firms and the innovation was available. Consequently 81 innovations constitute the sample of analysis.

25.4.1. Characteristics of the innovations awarded

The National Innovation Award has been organised in 2000, 2002, 2004 and 2006. Eighty-six, 198 and 166 applications were submitted for the first three editions. Most were sent by very small (less than 20 employees) and small (between 20 and 500 employees) construction firms. The most promising innovations were awarded (26 in 2000, 71 in 2002 and 57 in 2004)[3]. Innovations were classified into four categories by the organising committee: building techniques, safety and work conditions, methods and organisational schemes and environmental approaches.

Table 25.1 displays the number of applications per year and the number of awards per category[4]. These categories do not follow the concepts defined by the OECD and presented in the Section 25.2. Consequently, it was necessary to re-classify each innovation according to the guidelines provided by the Oslo Manual. For both the years 2000 and 2002, 43 innovations were given an award related to products (41 goods and 2 services); 17 to process innovations; 13 to organisational innovations and 8 were both a process and an organisational innovation. Table 25.2 displays the link between the type of innovation and the size of the firms.

25.4.2. The firms of the sample

Some of the size classes recommended by the Oslo Manual were used to classify the firms. The paper distinguished six categories: 1–9 employees, 10–49, 50–99, 100–249, 250–499 and groups (firms owned or controlled by a company such as Bouygues or Vinci). The firms from this last statistical unit (groups) may not have a large number

Table 25.2   Distribution of innovations according to firm size

| Size (employees) | Type of innovation | | | | |
	Product	Process	Organisational	Process and organisational	Total
1–9	12	6	0	0	18
10–49	11	6	4	3	24
50–99	0	0	1	2	3
100–249	4	1	3	2	10
250–499	1	0	0	0	1
Groups	15	4	5	1	25
Total	43	17	13	8	81

of employees but being an affiliate provides some advantages. For example, all large companies have an internal innovation awards' scheme such as the programme presented by Cousin (1998). These awards are open to all employees and concern all types of innovation. A technical director or a person in charge of promoting the innovations often helps the operatives who work on the building site to describe their approach and the innovation (the knowledge which is tacit becomes codified). This approach stimulates innovation and contributes to the creation of a true cultural value.

The firms from the sample are not representative of the population of French construction companies. They are larger but small companies are still dominant in the country as a whole[5]. Fourteen of these firms are suppliers (15 innovations). Two other companies were involved in the construction of infrastructure (mainly roads). Consequently, the analysis has been restricted to a sample of 64 innovations[6].

25.5. The role of the client in the innovation process

The client can be the source of the innovation and be involved during the course of the project. In the 64 cases analysed, the origin of the innovations is fourfold:

1. *The enterprise.* In these cases, the innovation results from the vision of manager or the information gathered by marketing people.
2. *Client(s).* The innovation results from the remark of a client who needs a better product or faces a specific problem.
3. *Regulations.* Empirical studies indicate that regulations can become one of the main drivers for innovation on the building site (Gann *et al.*, 1998; Pillemont, 2002; Bougrain, 2005). In three cases firms innovated to comply with the legal requirements established to prevent falls from a height. Twice it was due to the enforcement of the regulation concerning working time and enterprises modified their organisational scheme.
4. *A cost-oriented procurement process.* In four cases contractors were forced to innovate because they were involved in a construction project with a very low profit margin. Innovation was the only solution if the contractor was to have a profitable project (in three cases the client was a social housing company).

Table 25.3 Origin of innovations according to firm size

			Origin of innovations			
Size (employees)	The contractor	Client(s)	Regulations	Cost-oriented procurement process	Other	Total
1–9	7	1	1	1	2	12
10–49	16	1	2	0	0	19
50–99	0	1	0	1	0	2
100–249	4	1	0	2	0	7
250–499	0	0	0	0	0	0
Group	20	1	2	0	1	24
Total	47	5	5	4	3	64

Table 25.3 indicates that the role of the client is limited. Most innovations (73%) originated within the enterprise. This suggests that contractors undertake mainly project-specific innovations that improve the performance of the firm. As indicated in the Section 25.2, many innovations are ad hoc responses to bottlenecks that can jeopardise the realisation and the profitability of a construction project. In these cases, innovations originate within the enterprise.

Clients were at the origin of the innovation introduced by contractors on five occasions:

- In two cases builders of individual housing had to answer to the needs and the dissatisfaction of their clients. They were households who asked for houses with environmentally sensitive insulation materials and clients who had suffered from poor water tightness.
- One case concerned a contractor involved in demolishing a building insulated with asbestos. However the client, who demanded an innovative solution, was not involved in the resolution of the problem.
- One case concerned a service contractor who co-produced the solution with the client (a new disposal system for waste management).
- The last case dealt with a contractor specialising in heating, ventilation and air conditioning who had to find a solution to the ventilation problems encountered by a housing company. The client validated the solution later.

In the last two examples the clients also collaborated with the contractors and co-produced the final solution[7]. Their involvement predetermined the success of the project.

The contractor specialising in plumbing, heating, ventilating and air conditioning who employs 120 people, had to renovate the toilets of 8600 dwellings owned by a social housing company. The final goal of the client was to reduce water consumption. It was a cost-oriented procurement process. But the housing company had to cooperate to speed up the actions of the contractor and to limit the disturbances supported by the tenants. In this case the end user of the building (the tenant) had concerns regarding the project. During the project several meetings aimed at managing the implementation process, involved the specialised contractor, his equipment supplier and the social housing company. The supplier had to develop a new range of products in order to fasten the time spent by the plumbers on the building site and to solve the constraints linked to the age of the dwellings (traditional products did not fit to the dwellings that were more than 25 years old). Because of this win–win approach, the renovation was finished earlier than specified in the contract. Moreover, the project was profitable for the contractor and the supplier.

Such examples where the client is involved in the organisational process are not common. The majority of clients are not sophisticated and are rather passive. Clients are rarely involved because they are not expert in the problems encountered by contractors on the building site. Consequently they have no reason to interfere in the innovation process. They have more opportunity to be involved in the innovation process when they are not only the commissioners of the works but also the end users. In the six aforementioned cases, the clients were either the end users (e.g. households) or a representative of the users (e.g. housing companies). Even if the size of the sample is limited, it appears that clients have more chance to drive or at least take part in the innovation process when projects relate to their own business or when they are many issues at stakes.

25.6. Conclusion

After defining the notion of 'clients driving innovation', this chapter reports how clients may intervene in the innovation process led by contractors. The analysis mainly draws upon case studies of contractors recently rewarded for developing innovative products and methods. It indicates that clients do not often participate except in those cases where they are also the end user. In these cases they have an incentive to be involved because they will benefit from the innovation.

More empirical analyses need to be carried out to establish a typology of clients. For example, it would be interesting to analyse whether large public clients involved in public private partnerships, have a stronger influence on the innovation process. Under this new procurement process, once the design and construction is achieved, public clients have to occupy the building for more than 20 years. Undertaking case studies of operations within occupied buildings would also be interesting. Indeed in most industrialised countries, the management of the building stock (including maintenance and operating activities) is growing in importance in comparison with new construction. Consequently, opportunities to find innovative clients at this level would appear important.

Notes

[1] Large building companies, such as GTM, pioneered this approach by organising internal innovation awards in the 1990s. The aim is to disseminate best practices and new products/services within the group to avoid re-inventing the wheel next time. 'This process does not only enhance the innovation spirit, it also enables all responsible people, project managers and engineers, R&D specialists, to know of all innovations, to understand the

processes of innovation and the content of them, to transpose them into their own con-texts, to copy wherever possible, or to adapt them according to circumstances' (Cousin, 1998, p. 305).

[2] 'The limits of knowledge management techniques are not only driven by the project-based nature of activities, they also arise from high turnover, a reluctance on the part of engineers to recycle designs and incentive system within the profession, which rewards novelty rather than standardization' (Gann and Salter, 2000, p. 969).

[3] Innovations are awarded either at a regional or at a national level. At the national level, the jury also distinguishes the best 'Laureates'. In this empirical analysis we will not take into account these distinctions.

[4] Seventy-five per cent of the in-house innovations promoted by the GTM Awards scheme concerned improved methods (Cousin, 1998).

[5] In France, 92% of firms employ less than 10 employees. Forty per cent of the turnover of the sector is made by 1% of the enterprises. The average size of a construction company is 4.2 employees (Ministère des Transports, de l'Equipement, du Tourisme et de la Mer, 2006).

[6] However in some circumstances, the results obtained for suppliers are presented.

[7] Twenty-four of the 64 innovative projects led to the development of a technological collaboration.

References

Barrett, P. and Lee, A. (2005) *Revaluing Construction*. Rotterdam, CIB.

Bougrain, F. (2005) Strategies of roofing contractors who develop new equipments. *Proceedings of the 11th Joint CIB International Symposium: Combing Forces – Advancing Facilities Management and Construction through Innovation*, 13–16 June, Helsinki, Finland, p. 10.

Bougrain, F. (2003) Innovations in the building and construction industry: the case of material suppliers, manufacturers of building components and equipments and contractors. *Wirtschafts Politische Blätter* **3**: 366–371.

Brandon P. (2005) Vectors, visions and values—the essentials for innovation. In: *Clients Driving Construction Innovation—Mapping the Terrain* (eds K. Brown, K. Hampson and P. Brandon). Brisbane, CRC Construction Innovation, pp. 13–21.

Cousin, V. (1998) Innovation awards: a case study. *Building Research and Information* **26** (4): 302–310.

Drejer, I. (2004) Identifying innovation in surveys of services: a Schumpeterian perspective. *Research Policy* **33**: 551–562.

Egan, J. (1998) *Rethinking Construction*. London, Department of Trade and Industry.

Gallouj, F. (2002) *Innovation in the Service Economy: The New Wealth of Nations*. Cheltenham, Edward Elgar.

Gann, D.M. and Salter, A.J. (2000) Innovation in project-based, service-enhanced firms: the construction of complex products and systems. *Research Policy* **29**: 955–972.

Gann, D.M., Wang, Y. and Hawkins, R. (1998) Do regulations encourage innovation?—the case of energy efficiency in housing. *Building Research and Information* **26** (4): 280–296.

Hampson, K. and Brandon, P. (2004) *Construction 2020—A Vision for Australia's Property and Construction Industry*. Brisbane, CRC Construction Innovation.

Lundvall, B.-A. (1988) Innovation as an interactive process : from user-producer interaction to the national system of innovation. In: *Technical Change and Economic Theory* (eds G. Dosi, C. Freeman, R. Nelson, G. Silverberg and L. Soete). London, Pinter, pp. 349–369.

Manley, K. (2006) *Innovate Now! Improving Performance in the Building and Construction Industry. The BRITE Project.* Brisbane, CRC Construction Innovation.

Manley, K., Blayse, A. and Swainston, M. (2005) Implementing innovation on commercial building projects in Australia. In: *Clients Driving Construction Innovation—Mapping the Terrain* (eds K. Brown, K. Hampson and P. Brandon). Brisbane, CRC Construction Innovation, pp. 100–112.

Metcalfe, J.S. and Miles, I. (2000) *Innovation Systems in the Service Economy.* Boston, Kluwer Academic.

Ministère des Transports, de l'Equipement, du Tourisme et de la Mer (2006) Les enterprises de la construction en 2004. *SESP Infos rapides* (330): 4.

Mowery, D.C. and Rosenberg, N. (1979) The influence of market demand upon innovation: a critical review of some recent empirical studies. *Research Policy* **8**: 103–153.

National Audit Office (2001) *Managing the Relationship to Secure a Successful Partnership in PFI Projects.* HC 375 Session 2001–2002, 29 November, London.

OECD (2005) *Guidelines for Collecting and Interpreting Innovation Data.* Oslo Manual, OECD, European Commission, Paris.

Pavitt, K. (1984) Sectoral patterns of technical change: towards a taxonomy and a theory. *Research Policy* **13**: 343–373.

Pillemont, J. (2002) *Innovation, Qualité, Réglementation.* Paris, PUCA.

Sexton, M. and Barrett, P. (2003) Appropriate innovation in small construction firms. *Construction Management and Economics* **21**: 623–633.

Sirilli, G. and Evangelista, R. (1998) Technological innovation in services and manufacturing: results from Italian surveys. *Research Policy* **27**: 882–899.

Slaughter, S. (1993) Innovation and learning during implementation: a comparison of user and manufacturer innovations. *Research Policy* **22**: 81–95.

Sundbo, J. (1998) *The Organisation of Innovation in Services.* Frederiksberg, Roskilde University Press.

Winch, G. (1998) Zephyrs of creative destruction: understanding the management of innovation in construction. *Building Research and Information* **26** (5): 268–279.

26 A complex systems approach to customer co-innovation: a financial services case study

Robert Kay

26.1. Introduction

The topic of customer co-innovation is receiving considerably more attention in the literature of late, with authors such as von Hippel (2005) and Ulwick (2005) highlighting the role of customers in the product innovation process, despite the concept having been around for many years.

The notion of customer co-innovation highlights the limitations of non-customer based approaches to innovation by explaining many of the adoption failures associated with ideas or technologies that are developed with a customer in mind, but with an absence of customer input to the process of the idea's development. This is particularly the case in high technology contexts where highly specialised knowledge is assumed to be the pre-requisite for new thinking. Customer co-innovation is often conceived in these circumstances, in terms of the customer modifying the application of the idea or technology to fit their context specific purpose. Although the literature contains a number of studies that capture and discuss this aspect of customer co-innovation, they almost always refer to scenarios involving producers at one end and users at the other. What these studies ignore is the fact that users exist at all points along the value chain, and many innovations require co-innovation of the idea with multiple stakeholders in order for the idea to be adopted (Hendry, 2007). This is particularly the case in construction where large infrastructure projects span multiple domains of users and stakeholders, and in the financial services industry where the introduction of new payments-based technologies and processes have impacts not only at the point of the end consumer but at all points along the value chain (regulators, manufacturers, technology vendors, merchants, etc.).

The complexity created by this situation, makes the task of customer co-innovation less well defined than it is often described and requires new ways of looking at problems, and the process of innovation itself. From a strategic innovation perspective, these dynamics present an even greater problem: how does an organisation co-innovate with its customers around problems that might exist in the world 10 years from today?

This chapter will discuss an approach under development for the past 12 months within Westpac Banking Corporation in Australia. The approach differs considerably from those normally described within the literature, in that it focuses on co-innovation of what Baghai *et al.* (2000) describe as Horizon 3 type ideas. More specifically, it is focused on the Bank's corporate, as opposed to retail, customers and as such begins to deal with the value chain issues described above, by drawing together different elements of various value chains to explore and discuss the areas of commonality that they expect to have to deal with in the longer term (i.e. more than 10 years out from today).

The chapter first presents a description of the case study context, followed by the research methodology used. This is followed by the theoretical underpinnings of the approach and a general description of it.

26.2. Case study context

Westpac is Australia's oldest business, having begun in 1817 as the Bank of New South Wales. Renamed Westpac in 1982, the Bank is structured into four main divisions: 'Business and Consumer Banking' divisions (BFS and CFS), which provide retail banking services; Westpac Institutional Bank (WIB), which includes corporate, institutional banking and foreign exchange; BT Financial Group (BT), providing investment management and superannuation services and Business and Technology Solutions and Services (BTSS), which provides back-of-house processing and technology support to the other divisions. The Bank also has operations in New Zealand and a number of Pacific islands including Fiji.

Today, Westpac is the third largest by market capitalisation of what have become known as the 'big four' Banks in Australia. This concentration of large banks in Australia largely came about as a result of Government regulation known as the Four Pillars policy. The Four Pillars policy maintains the number of major banks by not allowing merger or acquisition, by overseas competitors or between the 'big four'. In effect, this situation has created a virtual oligopoly between the banks and although today the financial services market is experiencing increased competition in the form of overseas competitors entering the market and the emergence of other smaller Australian competitors, such as regional Australian banks, by and large the operating environment has been relatively benign and significantly stable. Indeed the services context itself often discourages attempts to innovate, due to the ability of competitors to rapidly copy any new service offerings that are brought to the market.

Despite this historical context, in October 2005, a small team was put together to explore what innovation could look like for Westpac. The team set about researching appropriate frameworks to inform the design and development of the innovation capability. Unsurprisingly, many of the characteristics of the Westpac approach to innovation can be found in other organisations. There is a stage-gated pipeline through which ideas are captured, assessed and developed through to commercialisation. The team reviewed the approaches of a number of different organisations from both within the services sector and also in manufacturing; copying, adapting and integrating different elements of these approaches into its own. Baghai *et al.*'s (2000) notion of strategic horizons was seen as a useful way of distinguishing between innovations at different stages of maturity. It is important to note, however, that the framework required considerable development beyond its description in Baghai *et al.*'s book, in order to be operationalised – this will be explained in more detail below. Future scenarios were also employed as a way to trigger ideas and think through different strategic contexts. The creation of filter committees to review ideas against specific filters was undertaken to assist in managing ideas as they came in. None of this is particularly innovative or different from the innovation approaches of a large number of organisations.

A point of difference in the Westpac approach, however, was the early recognition that the exploration of longer term opportunities was more effectively done with others, i.e. other organisations, as the creation of alternative futures, required as wide a range of perspectives as possible. This recognition then had logical implications for customer co-innovation, which will be discussed below.

26.3. Methodology

The research described in this chapter, was undertaken in the mode of a participant observation study, with myself as the researcher. As a member of the innovation team given responsibility for developing the innovation capability I was actively involved in the design and all decisions regarding the way in which the business case for innovation was developed, the methodologies applied and the ongoing evolution of the capability once it was funded.

26.3.1. Data collection

A reflective journal was maintained from the beginning of the initiative in October 2005 through to the writing of this chapter and will be ongoing for the life of my involvement in the programme. The journal catalogued key events and reflections under four broad categories:

- People met
- Decisions taken
- Theory examined
- Activities and events

The rationale behind these broad classifications developed over the initial months, moving through a couple of iterations. Overtime these classifications appeared to be the most common and as such the classifications stabilised around these four distinctions. The journal was updated on at least a weekly basis over this period, sometimes more often, depending upon activities taking place. The journal was maintained as an Endnote database with the fields renamed to match the above. The value of this approach is that it allowed easy searching of the journal for future reference and cross-referencing to reading material where appropriate.

26.3.2. Role of the researcher

My role in the team was to take principal responsibility for the design of ideation (idea generation) and modelling techniques, drawing specifically from the fields of complexity and systems theory in the way these processes were designed. I did not have ultimate responsibility for decision making for the innovation capability, this rested with the Head of Organisational Innovation and the organisation's Chief Technology Officer to whom the team reported. I, however, was actively involved in the discussions and always involved in the preparation of research material for them. As a participant researcher I had access to, and first-hand experience of, all aspects of the process and the documentation that emerged.

26.4. Customer co-innovation

Co-innovation with customers has long been practiced around the incremental end of the innovation continuum, indeed the many case studies we read always involve some new technology that has been developed, or a defined sub-culture of activity which over time has become defined as a market (e.g. Christensen, 2001; Ulwick, 2005; von Hippel, 2005). The team could not find examples of where customer co-innovation had been applied to the emergence of longer-term strategic ideas, i.e. those involving the reconfiguration of existing or indeed the creation of new value chains over the next 10–15 years. Furthermore, discussion of the types of methodologies that one would use in order to undertake this form of innovation also seemed absent.

Viewed through the lens of Baghai *et al.*'s (2000) '3 Horizon model', it became clear to the innovation team that the role of customers in the innovation process was not uniform to all ideas and could best be considered along a continuum. An idea's position on the continuum is dependent upon where the focus of the ideation process was aimed (Horizon 1, 2 or 3) and the stage of maturity of the idea. To understand why this is the case, its necessary to discuss Baghai *et al.*'s (2000) 3 Horizon framework in more detail. The 3 Horizon framework makes some very important distinctions about the nature of ideas or products/services at different stages of maturity. These distinctions are always made relative to the host organisation's existing strategic direction.

Horizon 1 refers to '[ideas] . . . that are critical to near term performance . . . continuing innovation can incrementally extend their growth and profitability . . . [however this growth] . . . will eventually flatten out and decline . . .' (Baghai *et al.*, 2000, p. 5). For Horizon 1 ideas, the range of flexibility and input by the customer is limited in scope by what is already known about the product. For example, what elements of the product can be changed and what cannot is likely to be known, the target demographic is likely to be largely understood, the underpinning technology has been tested, indeed the customer need can be observed first hand as they will already be solving it in some way.

Horizon 2 ideas or businesses (in Baghai *et al.*'s language) represent the next significant growth curve for a business. This is the typical space for customer co-innovation. Ideas at this stage of maturity, still require significant investment in order to capture their potential, with 'substantial profits maybe 4 or 5 years away' (Baghai *et al.*, 2000, p. 5) even though they may already be profitable. Intimate involvement of the customer as the idea continues to develop is crucial to the ongoing growth and evolution of the idea.

Horizon 3 ideas are the 'seeds of tomorrow's business', although characterised by real initiatives, the initiatives are always small and uncertain. 'A companies goal should be to keep the option to play without committing too much capital or other resources' (Baghai *et al.*, 2000, p. 7). From a customer co-innovation perspective, the focus is on a set of trends that provide a glimpse into the future, but little else will be known about the context. Not only does the customer not know what they need, but they also probably have no concept of the type of environment in which they might be living.

The creation of scenarios and simulation-based activities are the best opportunity to gain the customer's perspective, but that is all it will be as the real situation cannot be tested.

The distinctions made by the 3 Horizon model are significant as they neatly indicate the need for different forms of customer involvement on the basis of the characteristics of the ideas themselves. The model does have shortcomings, however, when it comes to operationalising these concepts. One of the difficulties relates to the range of different ways in which the authors conceptualise maturity. Sometimes the horizons refer to timelines, sometimes to geographies, sometimes profitability or size. This is OK when one is discussing a new growth business strategy built around a known product set. However, when attempting to look at idea maturity across a range of idea forms, a more consistent unit of analysis is required. Within Westpac, the team reinterpreted the horizons in terms of the degree of certainty held about the ideas in them. More specifically, the focus turned to understanding the patterns of behaviour that could be perceived in the environment of the idea. This reinterpretation is drawn from complexity theory on the basis that ideas seem to emerge and co-evolve with their environment over time. Certainly, the notion of customer co-innovation is a direct manifestation of, and enabler for, the co-evolution of an idea with its environment. Arguably the level of certainty held about an idea in Horizon 3 is quite low, relative to those in Horizon 1, but the question remained in regard to the methodological implications of this observation.

Kurtz and Snowden (2003), adopting a complex systems perspective, argue that the types of approaches applied to the resolution of perceived problems varies according to where the problem (idea) is located on their five space sense-making framework. Kurtz and Snowden's focus on problem resolution was consistent with the innovation team's approach to customer co-innovation, in the sense that their focus was on the identification of the 'problem to be solved' (Christensen and Raynor, 2003). Recognising that the level of certainty held in relation to the customer problem was the key unit of analysis allowed the team to begin mapping their initiatives relative to the concepts described in Kurtz and Snowden's framework.

26.4.1. Kurtz and Snowden's sense-making framework

The sense-making framework proposed by Kurtz and Snowden (2003) paints a picture of the patterns often associated with problems in an organisational context and links these patterns to the approaches that logically should be taken to solving them. The framework draws on ideas from the field of complexity science and as such is based on the assumption that patterns observed in nature can also be used to explain patterns of social behaviour.

Each domain in the framework represents a different form of dynamic, contextualised by the nature of the knowledge we have about that pattern. The simplest domain, called the 'Routine', refers to problematic patterns where the various parameters of the problem are *known* to its stakeholders and the implications of changing those parameters are also known. This is the domain of the simple problem. The next domain termed 'Complicated' refers to problems where the parameters associated with them are *knowable,* and can be solved through the application of appropriate expertise. For example, a lack of compatibility between two databases requires specific expertise to solve the problem, but essentially the scope of the problem is *known* even if the detail is only *knowable*. The third and most significant domain in the context of this chapter, Kurtz and Snowden (2003) calls the 'Complex'. Here the parameters associated with the problem can only be understood in retrospect. Often organisational and cultural change initiatives fall into this category as the complexity associated with the problems is so great that it is effectively impossible to understand the way in which the different parameters of the situation will interact before the event. Horizon 3 innovation typically falls into this category.

The fourth domain, Kurtz and Snowden (2003) term Chaos, where no patterns of cause and effect are discernible in relation to the problem. Generally speaking this is not a place you want to find yourself (although it could be debated whether finding yourself here is indeed possible). The last domain, which occurs at the intersection of the other four, is called 'Disorder', is in effect a space of non-agreement between stakeholders. This space is also significant in the innovation context as decisions in relation to whether a new idea should be supported or development stopped, regularly depend on the perception of the decision makers, above and beyond any rigorous framework of filters.

In the Westpac context, recognition of the differing organisational dynamics associated with different types of ideas has led to the development of parallel idea maturation/commercialisation processes. The reason for the parallel processes is that the degree of certainty an organisation has about an idea, changes its risk profile and consequently requires different positioning in terms of the exploration and exploitation processes in the organisation (March, 1991). Some of the implications of this point will be discussed below. However, it is worth noting at this point that the role of the customer depends very much on the type/form of innovation that is being considered.

26.5. Horizon 3 customer co-innovation

Within Westpac, the exploration of more typical forms of customer co-innovation were challenging for a number of structural reasons. Firstly, the innovation unit was situated in the technology and processing division of the Bank, as such it had no direct contact with the Bank's retail customers, nor a mandate to work with them. Secondly, as a function of this situation most of the ideas that moved through the innovation process were targeting either technology or process developments, rather than product development.

In a services industry, however, changes to processes have a direct impact on the experience of the customer and the way in which they receive their products. It quickly became clear that the development of *Complex* process-related ideas were best thought about in terms of the customer and in fact with the customer. The team were able to solve customer problems, not through the creation of products but through the co-innovation of the way in which they were received. This was particularly the case with corporate customers, where these changes had significant bottom line implications both for the client and for the bank.

Large corporate clients process billions of dollars worth of payments every year, giving rise to multi-million dollar relationships with the bank. Exploring the complete systems (both social and technical) associated with the payment processes of customers became a significant source of ideas and allowed the innovation team to create significant value for customers and facilitate improved client relationships for the institutional part of the bank. These initial successes, lead to further discoveries that helped the bank answer the question of how to co-innovate with customers around problems that were likely to occur long in the future.

26.6. Futures-based customer co-innovation

The innovation team had been working with scenarios (Schwarz, 1991; van de Heijden, 2005) since its inception as a way to help people think differently about their environment and generally broaden the ideation process. Overtime, the mode in which these futures were created fell within a strategic framework similar to what Johnston and Bate (2003) have termed frontiers. These frontiers were broad areas of opportunity that, according to the team's trend research, presented potential sources of growth. More broadly however, the frontiers represented areas of strategic interest to the organisation in terms of their longer term implications for the operating environment of the bank. In order to explore these areas of interest the team needed to significantly broaden the range of perspectives available to it, as put simply, the expertise in areas as diverse as genetically modified food through to bioinformatics and nanotechnology, were simply not to be found within the bank. This search led to the development of collaborative arrangements with a broad range of research-based organisations for which research in these diverse fields was their daily bread and butter.

What emerged from these collaborations was the identification of new pools of value that had not been identified before by any of the organisations involved. These value pools existed at the intersection between disciplines, industries, indeed the frontiers themselves. There were, of course, specific opportunities identified within the frontiers that although interesting and new, largely assumed an industry not radically dissimilar to that which already existed. But in terms of strategic innovation the focus moved to the spaces between which in essence represented the creation of new industry spaces emerging out of the interplay between the bank and its partners in the activity.

Recognition of the potential of this process was soon validated when some of these emergent ideas were explored with clients, but it was not the ideas themselves they were interested in, but the process of exploring new spaces at the points of intersection between industries. It was decided to extend the futuring process to become an activity that included the bank's corporate customers, who themselves had strategic frontiers that they perceived within and at the edges of their own industry. The combination of frontiers from both the bank and its clients in effect presented the first step in co-evolving or co-innovating the future. The process allowed the innovation team to co-generate the ideas with its customers in the context of longer term strategic staircases of development that the bank could enter into with the client and, in so doing, raising the nature of the relationship from a purely transaction-based delivery of service, to that of strategic partner in the real sense of the word.

26.7. Conclusion

There is little doubt that customer co-innovation and the involvement of lead-users in the innovation process significantly increase the chances of successfully commercialising new products and services. What this chapter has attempted to illustrate is that the nature of this co-innovation can be interpreted in a range of different ways, with significantly different outcomes for both the producer of the service and the customer. The approach described in this chapter, although representing the financial services context has obvious application in all industries and provides a means through which producers and customers can establish an ongoing context for co-innovation in and around longer term opportunities as well as more near term developments.

References

Baghai, M., Coley, S. and White, D. (2000) *The Alchemy of Growth: Practical Insights for Building the Enduring Enterprise*. New York, Basic Books.

Christensen, C. (2001) *The Innovator's Dilemma*. New York, Collins Business Essentials.

Christensen, C. and Raynor, M. (2003) *The Innovator's Solution: Creating & Sustaining Successful Growth*. Boston, Harvard Business School Press.

March, J. (1991) Exploration and exploitation in organizational learning. *Organization Science* **2** (1): 71–87.

Hendry, C. (2007) Innovation in fuel cells: dancing together or partners in tow? *Proceedings of the 23rd EGOS Colloquium*, 5–7 July, Vienna, Austria.

Johnston, R. and Bate, D. (2003) *The Power of Strategy Innovation: A New Way of Linking Creativity and Strategic Planning to Discover Great Business Opportunities*. New York, Amacom.

Kurtz, C. and Snowden, D. (2003) The new dynamics of strategy: sense-making in a complex and complicated world. *IBM Systems Journal* **42** (3): 462–483.

Schwarz, P. (1991) *The Art of the Long View*. New York, Doubleday Currency.

Ulwick, A. (2005) *What Customer's Want: Using Outcome-Based Innovation to Create Break-Through Products and Services*. New York, McGraw-Hill.

van de Heijden, K. (2005) *Scenarios: The Art of Strategic Conversation*, 2nd edn. Chichester, Wiley & Sons.

von Hippel, E. (2005) *Democratizing Innovation*. Massachusetts, MA, MIT Press.

Index